AF306654

Danilo Almeida Machado

Rarefied Flow Velocimetry

Danilo Almeida Machado

Rarefied Flow Velocimetry

Determining the speed of subsonic and supersonic free jets in an expansion chamber using the schlieren method

ScienciaScripts

Publisher:
Sciencia Scripts
is a trademark of
Dodo Books Indian Ocean Ltd. and OmniScriptum S.R.L publishing group

120 High Road, East Finchley, London, N2 9ED, United Kingdom
Str. Armeneasca 28/1, office 1, Chisinau MD-2012, Republic of Moldova, Europe
Printed at: see last page
ISBN: 978-620-8-01758-3

VELOCIMETRY OF RAREFIED FLOWS

Danilo Almeida Machado

1

DEDICATORY

This book is dedicated to all students and professionals in the field of optical diagnostics, fluid mechanics and thermodynamics. It has been written as a basic text, designed to provide fundamental knowledge and practical insights into velocity measurement in rarefied flows.

My hope is that this book will serve as a valuable resource, inspiring curiosity, enhancing understanding and promoting innovation within the community. Whether you are just starting out on your journey or are a seasoned professional, may this book contribute to your continued growth and success in this fascinating and essential field.

ACKNOWLEDGEMENTS

Firstly, I dedicate this book to God, whose guidance and blessings have been my constant source of strength and inspiration.

I would also like to express my deep gratitude to the friends I have made over the last 15 years during my academic experiences at São Paulo State University (UNESP), the Institute for Advanced Studies (IEAv), the Technological Institute of Aeronautics (ITA), the National Institute for Space Research (INPE) and the University of São Paulo (USP).

To my beloved family, I dedicate this work with sincere thanks. To my wife, Gabriela, your unwavering love and encouragement have been my anchor. To my son, Antonio, your curiosity and joy remind me daily of the importance of pursuing our passions. To my brother, José Mario, your constant support has been a pillar of strength, and in memory of my mother Renisia.

This book is a reflection of the collective wisdom, support and love I have received from each and every one of you. Thank you for being part of this journey.

SUMMARY

High-speed flows, whether they belong to the supersonic or hypersonic class, are systems intrinsically linked to technologies applicable to the aerospace sector. The ability to determine the velocity of a flow is particularly important if we consider its application as critical information in the validation of fluid dynamics computer codes. The aim of this work is to characterise the velocities of flows produced by an expansion chamber. To this end, three techniques were used to measure supersonic and subsonic velocities. An intrusive technique, using a Rapid Ionisation Detector, to characterise supersonic flows and two optical techniques, schlieren and a modified schlieren method combined with iodine molecular absorption. Using the Rapid Ionisation Detector, it was possible to study flows produced at a pressure of 10^{-6} mbar and with velocities ranging from 21 to 726 ms^{-1} . Using the method of characteristics to calculate the temperature and Mach number, it was possible to obtain flows with a minimum temperature of 29 K and a maximum Mach number of 12. With schlieren velocimetry it was possible to visualise flows under a vacuum of up to 200 mbar and with speeds ranging from 5.2 to 66.5 ms^{-1} . The flow temperature ranged from 300.0 to 297.6 K and the maximum Mach number was 0.2. By modifying the schlieren method, it was possible to visualise flows at pressures of up to 15 mbar, equivalent to an order of magnitude lower than the conventional schlieren method. The method measured flow velocities ranging from 92.0 to 190.0 ms^{-1} . The temperature ranged from 295.3 to 280.1 K and the maximum Mach number was 0.6.

Key words: Schlieren, Supersonic flow, Non-intrusive diagnosis.

List of symbols

τ	altura desobstruída da imagem
C_p	calor específico a pressão constante
C_v	calor específico a volume constante
k_{GD}	coeficiente de Gladstone-Dale
σ	coeficiente de viscosidade
q	comprimento característico do escoamento
λ	comprimento de onda da luz no meio
λ_0	comprimento de onda de luz no vácuo
k	constante de Boltzmann
C	contraste da imagem schlieren
ε_x	deflexão angular do raio na direção x
ε_y	deflexão angular do raio na direção y
ρ	densidade
ρ_0	densidade inicial
D	diâmetro do orifício da válvula de expansão
b	dimensão na direção x da fonte de luz
h	dimensão na direção y da fonte de luz
f_2	distância focal da segunda lente
f_3	distância focal da terceira lente
x	eixo das abscissas

y	eixo das ordenadas
z	eixo perpendicular aos eixos x e y
$H(x)$	entalpia molar do gás na posição x
L	extensão ao longo de um eixo óptico
n	índice de refração
n_0	índice de refração do meio circundante
Ω	fator de ampliação
Θ	livre caminho médio molecular
B	luminescência emitida pela fonte de luz
E	luminosidade de fundo
E_o	luminosidade incidente inicial
m	massa
K_n	número de Knudsen
M	número de Mach
M_{l2}	número de Mach do escoamento com sementes
$\dot{n}$	número de mols
d	operador da derivada
∂	operador de derivada parcial
A	parâmetro adimensional
x_0/D	parâmetro adimensional

ε	raios irregulares
δ	rarefação
γ	razão entre calor específico
S	sensibilidade da imagem schlieren
T	temperatura
T_{12}	Temperatura do escoamento com sementes
T_0	temperatura inicial
ΔE	variação da luminosidade
ΔE_{abs}	variação de luminescência causada pela absorção
u	velocidade
c_0	velocidade da luz no vácuo
c	velocidade da luz no meio
a	velocidade do som
u_{max}	velocidade máxima
u_m	velocidade média
$u(x)$	velocidade média do escoamento na posição x

Summary

1   Introduction

High-speed flows, whether they belong to the supersonic class, i.e. with a speed above the speed of sound, or the hypersonic class, at around five times the speed of sound, are systems intrinsically linked to technologies applicable to the aerospace sector. From the exhaust gases of an aeronautical turbine to the displacement of a launch vehicle in the Earth's atmosphere, the speed of a given flow is, alongside pressure, temperature, density and, in the case of reactive flows, the concentration of the species involved, an essential parameter for a complete characterisation of systems of this nature [1]. The ability to determine this set of parameters is particularly important if we consider its application as critical information in the validation of computational fluid dynamics codes, commonly known as CFD [2], *Computational Fluid Dynamics*, which are currently considered fundamental in all aircraft and space launch vehicle projects [3].

In the laboratory, the most convenient way to carry out research at high speeds is to make the gaseous medium flow over the object under study, which is at rest [4]. The devices usually used for this purpose are called shock tunnels, whose operation can produce supersonic or hypersonic flows s [5]. Shock tunnels consist of an extension of shock tubes, devices widely used in studies of the kinetics of chemical reactions at high pressures and temperatures [6]. A shock tube is a tube closed at its ends and separated internally by a diaphragm. In one section of the tube, a gas is placed at high pressure, while in the other part of the tube the study system is placed at low pressure. A shock tunnel is a shock tube with an expansion nozzle or tube separated by a second diaphragm added to the low-pressure end. There are currently three hypersonic shock tunnels in operation in Brazil, all located at the Institute for Advanced Studies (IEAv) [6]

The use of velocity sensors or any other type of probe designed to acquire and process information from the test section of shock tunnels comes up against the fact that these sensor elements act as shields to supersonic or hypersonic flow [7]. This obviously implies the creation of secondary shock waves, with the consequent disturbance of the hypersonic flow. As an alternative to using conventional sensors, optical methods, based on measurements associated with the emission, refraction, scattering and absorption of radiation, are a natural choice and are widely used in studies of hypersonic flows [,78].

Among the optical methods used to determine the velocity of flows in shock tunnels, the following stand out: i) Velocimetry by Rayleigh Scattering, which consists of the well-known relationship between the velocity of a wave-emitting particle and the Doppler effect associated with its displacement in space [9]; ii) Velocimetry by Filtered Rayleigh

Scattering, capable of instantly measuring the density, velocity and temperature of the flow. This makes it an attractive option, especially for high-speed or turbulent flows [10]. There is software developed for velocimetry in shock tunnels using Filtered Rayleigh Scattering. This software is similar to that used in Particle Image Velocimetry [11]; iii) Electron Beam Fluorescence Velocimetry, which is based on the excitation of atoms and molecules by the incidence of an electron beam [12]; iv) Particle Vaporisation Velocimetry, which is a technique in which the way the flow is marked is based on the vaporisation of particles by a laser at defined locations in the flow [13]; v) Velocimetry by Laser-Induced Fluorescence, which is based on the marking of a given chemical species present in the flow [14]; vi) Velocimetry by the Schlieren Optical Technique [15], through which it is possible to visualise the contrast variations in the flow image due to the variation in the refractive index associated with the homogeneities present along its direction of propagation.

In this study, the flows produced by a pulsed valve in an expansion chamber were characterised. The technique chosen was Velocimetry by Optical Schlieren Technique, used in its original version and in combination with molecular absorption spectroscopy. In addition, in some cases a FIG (*Fast Ionisation Gauge*) was used to determine the speed of the jets under investigation.

1.1 Schlieren method

The basic principle of the schlieren technique is to combine the optical projection of an object with an indication of its light deflection when propagating in heterogeneous media. There are various types of optical arrangements for the schlieren technique, which uses combinations of lenses and mirrors. Among them is the two-dimensional *Full* Field, which is the most established of the techniques [19] and which will be used in this work. Among the others are the *Rainbow* schlieren, which has recently been used for three-dimensional investigations [16]; the *Focusing* schlieren, with images similar to the Full Field, but with a simpler optical arrangement [17]; and the *Background* Oriented schlieren, which was developed for analysing images taken by digital cameras [18].

1.1.1 Light propagation in heterogeneous media

Light propagates uniformly through homogeneous media along a straight path. Starlight, for example, is hardly disturbed during its long journey through interstellar space. If the Earth's atmosphere were uniform, therefore, the starlight that hits us as parallel rays of light would not twinkle, and its brightness would be static, like that observed in space by telescopes and astronauts. However, the Earth's atmosphere is not homogeneous. On the contrary, it is made up of different layers of gases, and within the same layer, meteorological phenomena of a chemical and physical nature cause differences in temperature, pressure and density. The difference in density caused by these phenomena causes a variation in the refractive index of the medium, which in turn causes the light rays to be deflected from their axis of propagation as they pass through the heterogeneous medium. Once the wavefront is perpendicular to the axis of propagation, it compacts due to the phase change. The effect of this change is the twinkling of stars. Analogous reasoning cannot be applied to the light from planets, as they are not parallel [19].

Light slows down after interacting with matter. This phenomenon is expected by the Snell-Descartes law, where the refractive index (n) is given by:

$$n = \frac{c_0}{c} = \frac{\lambda_0}{\lambda} \qquad (1)$$

where c is the speed of light in the medium, c_0 is the speed of light in vacuum 3×10^8 ms^{-1}, λ is the length of light in the medium and λ_0 is the wavelength in a vacuum. In a gaseous medium, the relationship between density (ρ) and refractive index is given by:

$$n - 1 = k_{GD}\rho \qquad (2)$$

where k_{GD} is the Gladstone-Dale coefficient, equivalent to approximately 0.23 cm g$^{3-1}$ for air under normal temperature and pressure conditions, i.e. at 273.15 K and 101325 Pa and in visible light. For other gases, the coefficient can vary between 0.1 and 1.5 cm g$^{3-1}$. However, the refractive index of common gases only varies in the third or fourth decimal place. The parameter k_{GD} decreases slightly as the wavelength of light increases. Therefore, the refractivity is greater for smaller λ, so the variations in the refractive index are more intense for ultraviolet rays than in visible light.

The refractivity of a gas depends on the composition of the gas, temperature, density and the wavelength of the incident radiation (λ). These quantities can be related by the equation of state of the perfect gas.

$$(n - 1) = \frac{k_{GD}\,pm}{\dot{n}RT} \qquad (3)$$

where T is the temperature, m is the mass of the gas and $\dot{n}$ is the number of moles. Considering any gas, all variations in temperature, density and pressure lead to disturbances in the gas. As light passes through heterogeneous regions, its path is altered as a result of the variation in refractive index.

From the point of view of geometric optics, the deviation of light rays caused by heterogeneities in the medium can be presented as follows:

$$\frac{\partial^2 x}{\partial z^2} = \frac{1}{n}\frac{\partial n}{\partial x}; \qquad\qquad \frac{\partial^2 y}{\partial z^2} = \frac{1}{n}\frac{\partial n}{\partial y} \qquad (4)$$

where the x and y axes are perpendicular to the normal z direction.

Integrating the deviation of the light rays to first order, we have the angular deflection components of the rays (ε) in the x and y directions:

$$\varepsilon_x = \frac{1}{n}\int \frac{\partial n}{\partial x}\partial z; \qquad\qquad \varepsilon_y = \frac{1}{n}\int \frac{\partial n}{\partial y}\partial z \qquad (5)$$

For a two-dimensional analysis of length L along an optical axis, the equation for the angular deflection components is as follows:

$$\varepsilon_x = \frac{L}{n_0}\frac{\partial n}{\partial x}; \qquad\qquad \varepsilon_y = \frac{L}{n_0}\frac{\partial n}{\partial y} \qquad (6)$$

where n_0 is the refractive index of the surrounding medium.

1.1.2 Theoretical foundation of the Schlieren Toepler Total Field Optical System

In a full-field schlieren optical array it is possible to visualise the entire length of an object that falls within the system's test region. Figure 1 shows an example of a full-field Toepler schlieren optical setup. The lenses L_1 and L_2 function as a beam expander. The path between lenses L_2 and L_3 consists of the test region, where the light beam is parallel. When it passes through lens L_3, the light is converged at the point where a spatial filter, known in technical jargon as a knife, is located. When the refracted light is deflected by a variation in the refractive index in the test area, it is obstructed by the knife. This causes the image formed on the camera to be a set of points with a contrast in light intensity. In other words, the difference in phase between the light rays has been converted into a difference in brightness, easily recorded by the available instrumentation.

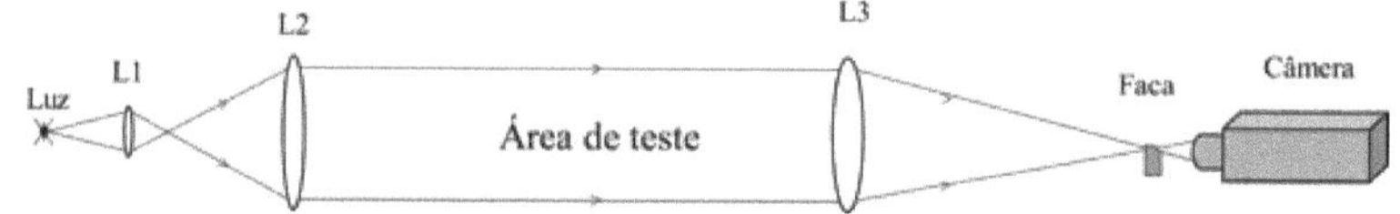

Figure 1 - Full Field schlieren optical arrangement, L_1, L_2 and L_3 are converging lenses

The sensitivity of a measuring instrument is one of its basic characteristics and is associated with its level of response to the information received at the input. In the case of full-field schlieren optics, the output is a two-dimensional image. Figure 2 shows a plane with an infinite depth field, where the x and y axes form an image and the z axis is considered to be the path of the light rays. More specifically, the output consists of a set of image elements characterised by contrast variations.

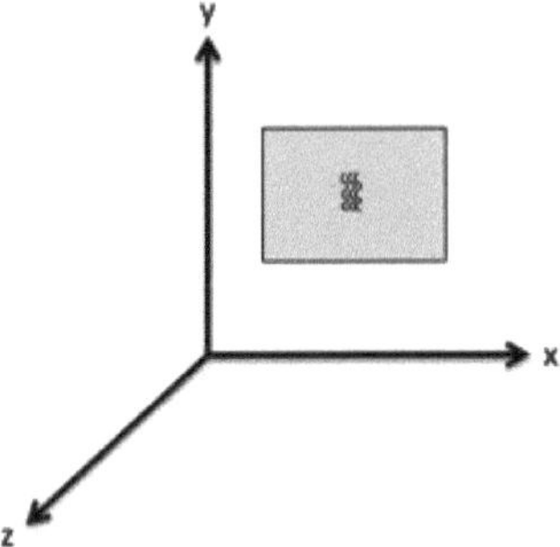

Figure 2 - The output of the schlieren image corresponds to an (x, y) plane.

The luminosity E_o incident on the lens L_2 can be calculated using the inverse square law, where the luminosity decreases with the square of the distance, given by:

$$E_0 = \frac{Bbh}{f_2^2} \qquad (7)$$

where b and h are the width and height dimensions of the light source, B is the luminescence emitted by the light source and f_2 is the focal length of the lens L_2 . Considering a magnification factor Ω that represents the size of the image in relation to the test area, we have:

$$E_0 = \frac{Bbh}{\Omega^2 f_2^2} \qquad (8)$$

In the case of the luminosity incident on L3, the focus of the system is given by the product of the foci of lenses L2 and L3, while the parameter

τ replaces h, which is the unobstructed height of the image formed after the spatial filter.

$$E = \frac{Bb\tau}{\Omega^2 f_2 f_3} \qquad (9)$$

Each schlieren image has the background illumination E. The illumination of a point in the image is evaluated by comparison with this intensity level: higher, lower, or equal to it.

Considering the case of an object in the test area of the schlieren arrangement, the light rays from the source will undergo a deviation of ε, where the y component of the angle of refraction is ε_y . Since the variation in image height is given by the product of the angle of refraction and the focal length of the lens L3, we have:

$$\Delta E = \frac{Bb\varepsilon_y}{\Omega^2 f_3} \qquad (10)$$

The contrast in the schlieren image, which is the ratio of the change in luminescence ΔE to the background luminescence E, is:

$$C \equiv \frac{\Delta E}{E} = \frac{f_3 \varepsilon_y}{\tau} \qquad (11)$$

The contrast in the image corresponds to the output of the schlieren arrangement. The input is a pattern of irregular rays of angle ε resulting from the refractive index gradients in the test area. Since the schlieren sensitivity (S) is basically the ratio between the output and input of the light in the system, it follows that:

$$S = \frac{dC}{d\varepsilon} = \frac{f_3}{\tau} \qquad (12)$$

This simple result of geometric optics provides a measure of schlieren sensitivity independent of any observation.

1.1.3 Contrasting schlieren technique

To discuss the parameters that determine the contrast of the schlieren technique, let's imagine a hypothetical condition in which the flow is produced from the pressure difference between two regions. This flow condition is the same as that produced by shock tubes and tunnels and valves. In this hypothetical situation, the contrast of the schlieren image obtained by a camera is related to 8 parameters, 3 of which refer to the flow and 5 to the optical instruments used in the experimental setup.

(i)   The refractive index of the gas in the flow is a parameter that is related to the deflection of the light rays, according to equation (6). For higher refractive index values, the schlieren contrast will be more intense.

(ii)   For increases in the density, pressure and temperature of the flow, the schlieren contrast will be more intense.

(iii)   For increases in the density, pressure and temperature of the background gas against which the flow has expanded, the schlieren contrast will be more intense.

(iv)   The wavelength of the light source in the schlieren arrangement is related to the Gladstone-Dale coefficient according to equation (2). For higher coefficient values, the schlieren contrast will be more intense.

(v)   For increases in the intensity of the light source of the schlieren arrangement, the schlieren contrast decreases; these two parameters are related according to equation (11).

(vi)   The focal length of the lens or mirror that captures the parallel light that interacts with the flow is related to the sensitivity according to equation (12). For larger focal length values, the schlieren contrast will be more intense.

(vii)   In equation (11), the $\mathit{\Delta E}$ term is a function of the percentage cut of the rays refracted by the knife. Some works in the literature indicate that the minimum cut to achieve the schlieren effect is between 5% and 10% [19,20]. The percentage of cut rays that provide the best contrast varies according to experimental settings and the view of the observer.

(viii) Schlieren contrast also depends on the sensitivity of the camera capturing the images. The more sensitive the camera's detector, the lower the intensity of the light source, item (v), and consequently the higher the schlieren contrast.

Table 1 summarises the discussion of this sub-item, showing the dependence of the contrast on the flow parameters and the optical arrangement.

Table 1 - Contrast dependence of optical and flow parameters.

Increase in optical and flow parameters	Contrast	
	increases	decreases
Refractive index	x	
Density, pressure and temperature of the flow	x	
Density, pressure and temperature of the background gas	x	
Wavelength of the light source		x
Intensity of the light source		x
Focal length	x	
Camera Sensitivity	x	

1.1.4 Space applications

Our eyes and ordinary cameras cannot discern the phase differences in a beam of light. We can only see the amplitude and colour contrast. If it were possible to see the phase of the light, new perspectives would open up for our vision. Such is the importance of the schlieren optical method, it translates phase differences into amplitude [19].

The father of optics in non-homogeneous media was Robert Hooke. His fascination with atmospheric refraction led him to establish the optics of heterogeneous media as a new field of scientific enquiry. In 1672, in an experiment involving lenses and a candle flame, Hooke visualised disturbances in the air near the flame. This experiment sowed the seeds for the development of the schlieren optical technique [21]. However, the first to officially recognise and present the schlieren imaging technique was August Toepler, who invented a functional apparatus for generating schlieren images, as well as a detailed procedure on how to design a simple schlieren system [22]. Since then, the technique has been used mainly to visualise flows.

Figures 3, 4 and 5 are examples of the study of flow in space applications using the Total Field Schlieren technique. The images were taken in the laboratories of the Aerothermodynamics and Hypersonics Division of the Institute for Advanced Studies and represent, respectively, some of the studies in the hypersonics laboratories: study of energy addition in flows to reduce drag [23]; study of SCRAMJET *Supersonic Combusting Ramjet* engines [24]; study of the SARA (Atmospheric Reentry Satellite) [25] and study of laser propulsion [26].

Figure 3 (a) and (b) were obtained in the T2 shock tunnel, both without the presence of a flow. Figure 3 (a) corresponds to the image of the plasma formed by a laser focussed on the test region of the tunnel, the image was taken 120 µs after the laser pulse, under atmospheric pressure conditions. Figure 3 (b) corresponds to a pulse of the same power fired in an environment with a pressure of 60 torr.

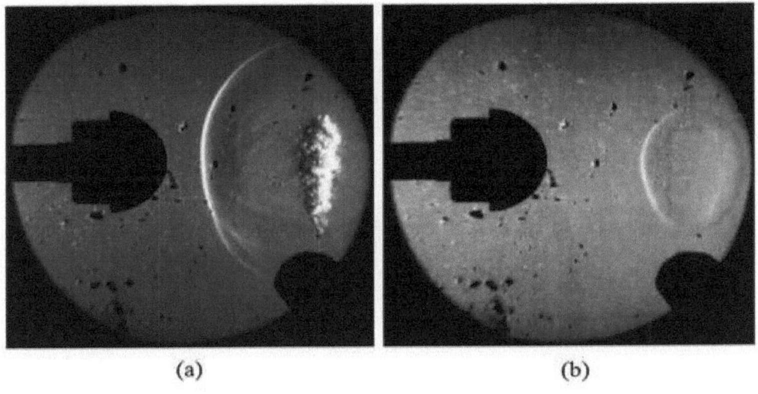

Figure 3 - Schlieren photographs of laser energy addition in static medium for gas pressure: (a) atmospheric pressure, (b) 60 torr. [5]

Figure 4 corresponds to a schlieren image of the inlet of a SCRAMJET engine taken in the T3 shock tunnel. The atmospheric air shockwave enters the compression ramp where hydrogen gas is injected, Figure 4 (a). When it comes into contact with hydrogen, the oxygen present in the shock wave starts supersonic combustion, Figure 4 (b).

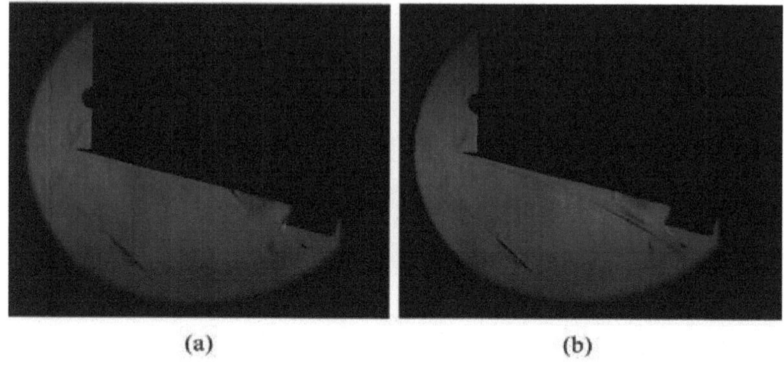

Figure 4 - Schlieren photographs of a SCRAMJET ramp, (a) hydrogen injection, (b) hydrogen compression by the shock wave. [24]

Figures 5 (a) and (b) are images from the SARA study. The photographs were taken in the T2 tunnel. Image 5 (a) corresponds to a conventional time-integrated photograph of the plasma formation of a shock wave of atmospheric air coming into contact with the surface of the model. Image 5 (b) is a schlieren photograph from another SARA model. The formation of the shock wave over the model can be clearly seen.

Common to all the application examples presented is the production of heterogeneous regions along the flow direction, from which it is possible to obtain contrast in the schlieren image. This phenomenon is very characteristic of spatial systems, which makes the schlieren technique one of the most immediate and complete research options.

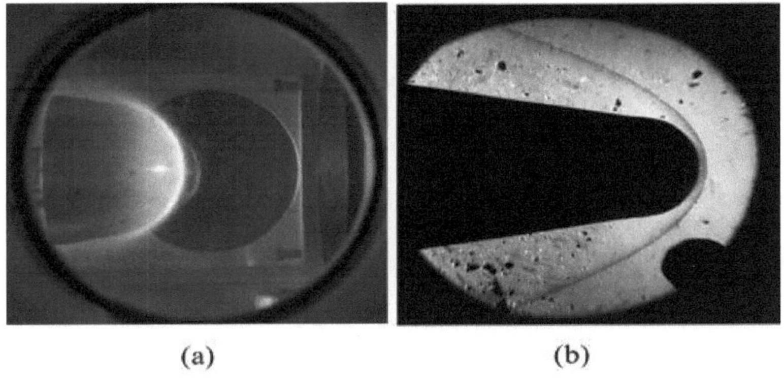

(a)   (b)

Figure 5 - Photographs from the SARA study, (a) conventional photograph, (b) schlieren photograph. [25]

1.2 Schlieren Optical Method Combined with Molecular Absorption (SCAM)

Visualising a flow using the schlieren optical technique is possible thanks to the deviation caused by variations in the pressure, temperature and density gradient. The knife has the function of cutting the refracted light, and the image formed is given by the contrast of the light as it passes through the medium containing the flow. The greater the deviation of the rays as they pass through the test area, the greater the contrast of the image. In certain physical conditions of the medium, such as low temperature, pressure or density, the refraction of light does not generate contrast. Under these conditions, it is not possible to visualise the flow using the conventional optical schlieren method. In order to increase the contrast, a modification of the method was proposed in this work.

In addition to variations in the conditions of the medium, contrast can be increased by using absorbing species seeded in the medium under investigation. To do this, molecules that absorb at the wavelength of the light source are introduced into the flowing gas. In this case, the molecules act as a shield to the passage of light, increasing the contrast in the image. Consequently, the Schlieren Combined with Molecular Absorption (SCAM) optical method is capable of investigating flows produced in certain physical conditions that the conventional method cannot. In other words, it has greater sensitivity.

Figure 6 illustrates the effect on image contrast. The solid line represents the light rays that have not been refracted, the dotted line represents the light rays that have been refracted and the dashed line represents the light rays that will be absorbed by the seeds contained in the gas. In Figure 6 (a) the contrast in the image corresponds to a balance of the refracted rays and the background light. In Figure 6 (b) the contrast in the image corresponds to a balance of refracted rays, absorbed rays and background light.

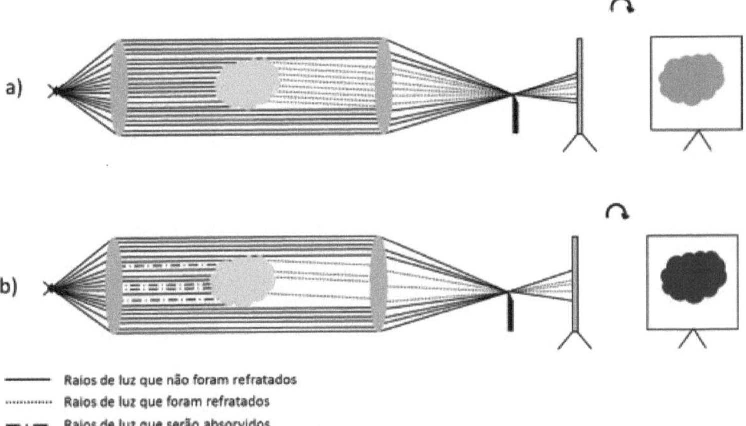

Figure 6 - Figure (a) corresponds to conventional schlieren, where the contrast is given by the refracted rays. In Figure (b), the contrast is given by the combination of molecular absorption and refracted rays.

From a mathematical point of view, the light rays that are absorbed contribute as an increase in the ΔE parameter of equation (11). The variation in luminescence ΔE corresponds to the balance of refracted rays. With the addition of absorbing molecules, equation (11) gains a new term ΔE_{abs} which corresponds to a variation in luminescence caused by the absorbed rays.

$$C \equiv \frac{(\Delta E + \Delta E_{abs})}{E} \qquad (13)$$

1.3   Expansion chamber

Wind tunnel measurements are expensive and complex. For example, the cost of operating IEAv's T3 is $c.$US\$3,000.00 per test, and the maximum number of tests is one per day. For the development of optical techniques, it is desirable to have a system in which a greater number of tests can be carried out in order to establish the technique. In this sense, the use of a vacuum chamber equipped with a pulsed valve can simulate conditions similar to those in shock tunnels [,2728].

Flows produced in a vacuum chamber can reach a velocity of *around* 800 ms^{-1} and a Mach number of *around* 30 [29]. Some research groups have had great success in developing optical techniques in expansion chambers. M. M. Koochesfahani's group at Michigan State University, for example, has developed an optical technique for measuring flow pressure based on the molecular labelling of acetone [30]. S. W. North's group at the University of Texas has developed an optical technique capable of simultaneously measuring the temperature and velocity of a flow through the excitation of vibrational states of nitric oxide [31]. In this work, the schlieren optical technique for measuring flow velocity was established in an expansion chamber designed and built at the IEAv.

An expansion chamber, Figure 7, consists of a device basically made up of an expansion valve, a vacuum system, an exhaust valve and windows for optical measurements.

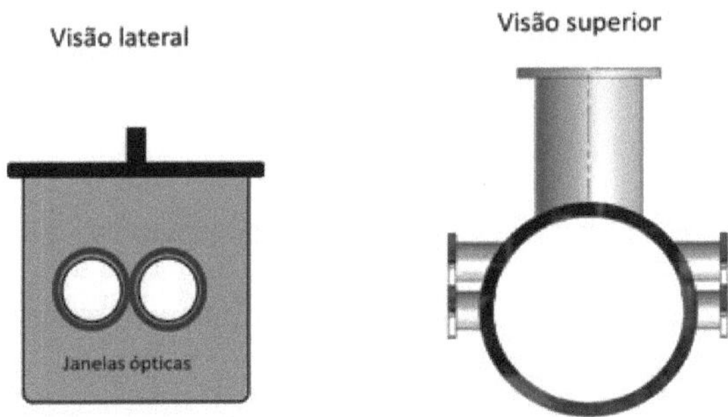

Figure 7 - Illustration of an expansion chamber.

The flow velocity in the expansion chamber can be controlled from the relationship between the exhaust valve stagnation pressure and the expansion chamber pressure. By reducing the chamber pressure, it is possible to obtain a higher flow velocity. Using the theory of Rarefied Gas Dynamics, it is possible to predict the flow regime produced by the operation of the pulsed valve system in the expansion chamber, as shown in the next section.

1.4 Rarefied Gas Dynamics

The main parameter used in Rarefied Gas Dynamics is the Knudsen number [32], (K_n), defined as the ratio between the mean free molecular path, ($\odot$), and a characteristic length, (q), of the flow. For the expansion chamber system using a pulsed valve, the characteristic length (q) corresponds to the diameter of the valve orifice.

$$K_n = \frac{\odot}{q} \qquad (14)$$

In practical calculations, the rarefaction parameter (δ) is used, which is inversely proportional to the Knudsen number. The mean free path depends on the intermolecular interaction potential whose parameters are determined using the viscosity coefficient. Rarefaction is given by:

$$\delta = \frac{\sqrt{\pi}}{2} \frac{1}{K_n} = \frac{pq}{\sigma u_m} \qquad (15)$$

where σ is the viscosity coefficient and u_m is the average flow velocity. From the rarefaction calculation we can divide the gaseous flow regime into three types [33]:

(i)   Regime of free molecules, where $\odot \gg q$. In this regime, the mean free path is much longer than the characteristic length of the flow. The interaction between the particles can be disregarded, i.e. in this regime the particles are considered to be moving independently of each other. In an expansion chamber, this is the necessary condition to produce *Supersonic Molecular Beams* (SMBs) [34].

(ii)   Transitional regime, in which $\odot \approx q$. In this regime, the interaction between particles cannot be disregarded as in the free molecule regime, nor can the medium be considered a continuum.

(iii)   Hydrodynamic regime, where $\odot \ll q$. In this regime the mean free path is much smaller than the characteristic length of the flow and, consequently, the gaseous medium can be considered as a continuous medium in which the equations of hydrodynamics can be applied.

1.4.1 Continuous flow

The most basic way of describing a theory for modelling continuous flows was first stated in 1951 by A. Kantrowitze and J. Grey [35]. An experimental proof of the fundamental principles was demonstrated shortly afterwards by G.B. Kistiakowsky and W.P. Slichter [ ]. Slichter [36]. The equations developed to describe the flow parameters of interest such as temperature, velocity, density and Mach number are similar to those used in fluid dynamics for aircraft design s [37].

As a boundary condition for the equations, in the description of continuous flow, the gas is considered ideal. In addition, the flow is considered to be an adiabatic and isentropic system, i.e. its entropy is constant, there is no heat exchange and the process is reversible. Under these conditions, the effect of viscosity in the gas can be neglected. Considering a system in which the diameter of the expansion valve orifice (D) is much larger than the mean free path of the molecules (λ_o), ($D \gg$ $\odot$)the adiabatic assumption leads to the conservation of the sum of the enthalpy and kinetic energy of the mass of the flow in a vacuum expansion [37], so that:

$$H(x) + \frac{1}{2}mu(x) = constante \qquad (16)$$

where $H(x)$ is the molar enthalpy of the gas at position x, m is the mass of the gas and $u(x)$ is the average velocity of the flow at position x. The maximum speed that a freely expanding flow can reach is given by:

$$u_{máx} = \left(\frac{2H(T_0)}{m}\right)^{1/2} \qquad (17)$$

where $H(T_0$) is the molar enthalpy of the gas as a function of the temperature of the gas reservoir. For a thermodynamic system in which the pressure is constant, the specific heat C_p is independent of the temperature and its enthalpy variation is directly proportional to the variation in heat received by the system.

$$dH = C_p dT \qquad (18)$$

In general, the hypothesis of an adiabatic isentropic expansion is reasonably well validated by means of temperature measurements near the outlet of the flow [38]. By substituting equation (18) into (17) we have a measurable relationship for the maximum velocity.

$$u_{máx} = \left(\frac{2C_p T_0}{m}\right)^{1/2} \qquad (19)$$

The initial stagnation conditions (p_0, T, ρ_{00}) and final conditions (p_1, T, ρ_{11}) of an adiabatic isentropic expansion are related by the following equations:

$$\frac{T_1}{T_0} = \left(\frac{p_1}{p_0}\right)^{\gamma-1/\gamma} ; \qquad \frac{\rho_1}{\rho_0} = \left(\frac{p_1}{p_0}\right)^{1/\gamma} ; \qquad \frac{\rho_1}{\rho_0} = \left(\frac{T_1}{T_0}\right)^{1/\gamma-1} \qquad (20)$$

where p is the gas pressure, ρ its density and γ is the ratio between specific heat at constant pressure C_p and specific heat at constant volume C_v, given by:

$$\gamma = \frac{C_p}{C_v} \qquad (21)$$

The propagation of the speed of sound in a gas flow can be considered an adiabatic and quasi-isentropic process [39]. Propagation takes place through a variation in pressure caused by the passage of the sound wave. However, the speed of sound does not depend on pressure, but increases with the square root of the absolute temperature [40].

$$a = \left(\frac{\gamma kT}{m}\right)^{1/2} \qquad (22)$$

The Mach number, M, is a dimensionless quantity that is defined as the ratio between the speed of the flow and the speed of sound.

$$M = \frac{u}{a} \qquad (23)$$

One of the physical interpretations for the Mach number is to make it the direct measure of the movement of the gas compared to the random thermal movement of the molecules, since this parameter considers the flow moving along one direction.

1.4.2 Supersonic molecular jets

The Supersonic Molecular Jet is a flow regime in which molecules, usually in the range of 0.7 to 100 atmospheres, are expanded through a small orifice into a region of low pressure $ca.$ 10^{-6} torr [41].

Figure 8 illustrates an expansion of an SMB in which two regions are observed. The collision region occurs close to the nozzle. As the gas expands, the molecular region is formed, where little or no collision occurs.

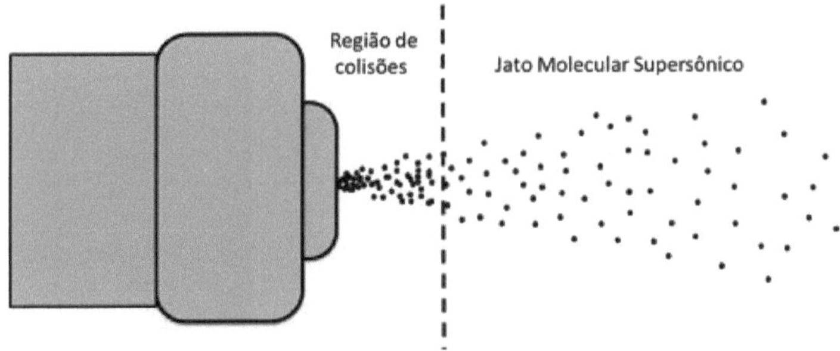

Figure 8 - Free expansion of a valve divided into two regions.

In the region where molecules collide, the rotational, vibrational and transverse kinetic energies of the molecules are converted into kinetic energy in the direction of flow [42]. This process causes the relative velocity between the molecules to decrease, which in turn causes a decrease in temperature. When travelling at a low relative speed and close to each other, the molecules feel an attractive force called the Van der Waals force, capable of causing the formation of particle clusters [43].

The most efficient method of producing SMBs is through the use of pulsed valves in vacuum chambers. The first valves for this purpose were developed in 1978 by Gentry and Giese [44]. The system they created was based on the principle of magnetic repulsion between metal strips generated by a pulsed electric current. In 1982, the pulsed piezoelectric valve was developed by Jon Cross and James Valentini [45]. This valve operates using piezoelectric crystals which, when subjected to a pressure difference, generate an electric field.

Pulsed SMB has the advantage of having a lower density than continuous SMB. The dynamic cooling process is more efficient, making it possible to study lower rotational states. With the use of a *skimmer*, which works like a collimator for supersonic beams, a smaller range of velocity distributions can be selected [46]. However, the biggest advantage of working with pulsed valves is the possibility of operating with a lower pumping rate of the vacuum system, due to the smaller volume of gas inserted by the valve.

The study of SMBs has a wide range of applications, including: the study of rocket engines and satellites that use valves as a propulsion system [47]; reduction of cluster formation for laser isotope separation [48]; low temperature SMBs allow the study of certain ro-vibrational transitions in Laser Induced Fluorescence spectroscopy [49]; recent study of temporally and spatially resolved measurements of plasmas in SMBs show a direct relationship between the plasma's electric field intensity and its velocity [50].

1.4.3 Characteristics method

In 1948, the Method of Characteristics for SMBs was first proposed by P. L. Owen and C. K. Thornhill [51], and later developed by H. Ashkemas and F. S. Sherman [52]. In this method, the SMB gas is considered ideal and non-viscous in a one-dimensional, adiabatic and isentropic flow [52]. The Characteristics Method describes the continuous regime and the change from the continuous regime to the molecular regime. For the molecular regime, the flow molecules may not be in thermodynamic equilibrium. Normally, the rotational mode reaches equilibrium faster than the vibrational mode, so vibrational equilibrium may not occur during expansion [53].

According to this method, the Mach number can be obtained theoretically as a function of the distance from the valve outlet, given by:

$$M = A \left(\frac{x_d - x_0}{D}\right)^{\gamma-1} - \frac{\frac{1}{2}\left(\frac{\gamma+1}{\gamma-1}\right)}{A\left(\frac{x-x_0}{D}\right)^{\gamma-1}} \qquad (24)$$

where the terms of the equations for A and x_0 /D as a function of the ratio of the specific heats are obtained from the reference [54]:

$$A(\gamma) = 211\gamma^4 - 1170\gamma^3 + 2420\gamma^2 - 2220\gamma - 7680 \qquad (25)$$

$$\frac{x_0}{D}(\gamma) = 115e^{-3,94\gamma} \qquad (26)$$

where x_d is the distance from the valve orifice on the flow axis and x_0 is the start of supersonic expansion on the flow axis. Table 2 shows the values of the constants in equations (25) and (26) calculated from different specific heat ratio values.

Table 2 - Values of the constants.

γ	x/D_0	A
1,67	0,075	3,26
1,40	0,400	3,65
1,30	0,700	3,90
1,29	0,850	3,96
1,20	1,000	4,29
1,10	1,600	5,25
1,05	1,800	6,44

For diatomic molecules, the ratio of specific heats can be considered constant over the temperature range from 3 to 600 K, *with* a value of 1.40 [55]. By applying this value to equation (24), the Mach number found in the beam can be simulated. Figure 9 shows the Mach number as a function of distance from the hole.

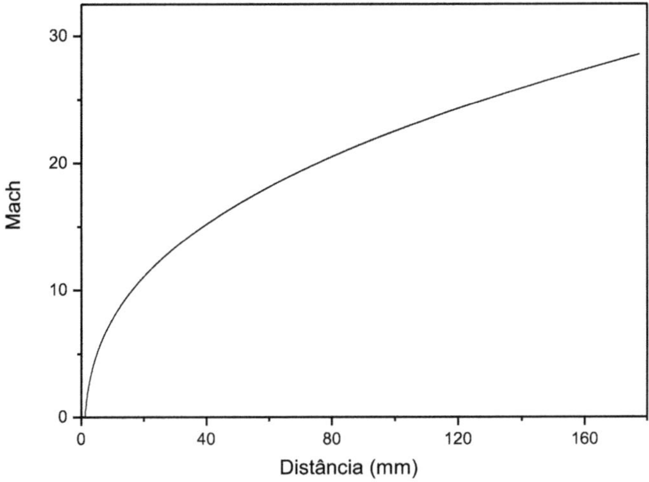

Figure 9 - Mach number as a function of distance, calculated from equation (24)

for diatomic molecules.

The temperature of the SMB can be obtained as a function of the Mach number, the ratio of the specific heats and its initial temperature. The initial temperature is taken to be the same as the ambient temperature.

$$T = \frac{T_0}{1 + (\gamma - 1)\frac{M^2}{2}} \qquad (27)$$

Figure 10 shows temperature as a function of distance from the valve. In a flow, a relatively common mistake is to establish a direct relationship between temperature and velocity. However, in a flow, the concept of gas temperature represents the average kinetic energy of the particles in relation to the centre of mass of the flow and not in relation to the laboratory reference frame. In the process of dynamic cooling, the movement of the particles in the reference frame of the centre of mass of the flow tends to be zero, causing an exponential decay in temperature.

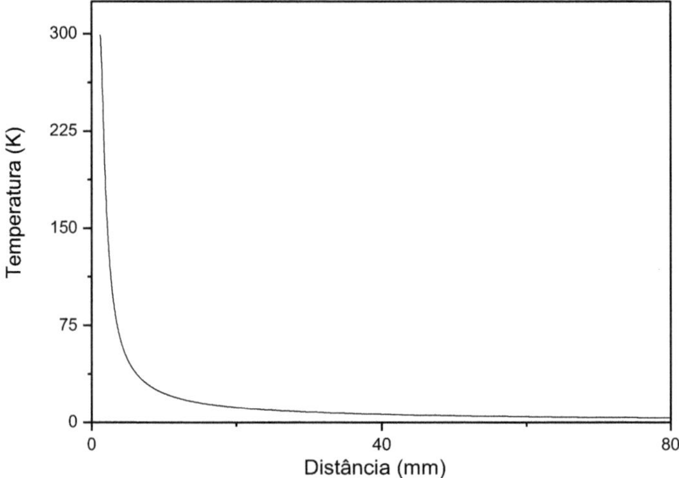

Figure 10 - Temperature as a function of distance from the valve calculated from equation (27) for diatomic molecules.

The density of the beam can be calculated as a function of the Mach number, the ratio of the specific heats and its initial density given by:

$$\rho_0 = pm/nRT_0 \qquad (28)$$

So the density is:

$$\rho = \frac{pm/nRT_0}{\left[1 + (\gamma - 1)\frac{M^2}{2}\right]^{1/\gamma - 1}} \qquad (29)$$

By substituting equations (24) and (27) into (23), the SMB speed can be obtained as a function only of the constants and the distance from the beam to the valve orifice (x) expressed in equation 29. The square root of the bracket represents the speed of sound.

$$u = M \left[\frac{\gamma RT_0}{1 + (\gamma - 1)\frac{M^2}{2}}\right]^{\frac{1}{2}} \qquad (30)$$

Figure 11 shows the simulation of speed as a function of distance from the valve. It can be seen that the speed increases sharply as a function of the distance from the orifice, remaining constant from 10 mm onwards.

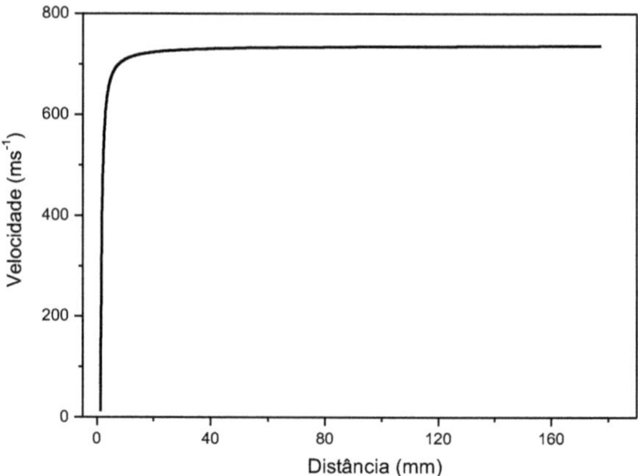

Figure 11 - Velocity as a function of distance from the valve, calculated from equation (30) for diatomic molecules.

Figure 12 shows the simulation of temperature as a function of velocity for supersonic molecular beams for diatomic molecules. As the temperature decreases, the flow velocity increases. The maximum speed that the beam of diatomic molecules can reach is *ca.* 740 m/s, and its minimum temperature is *ca.* 12 K.

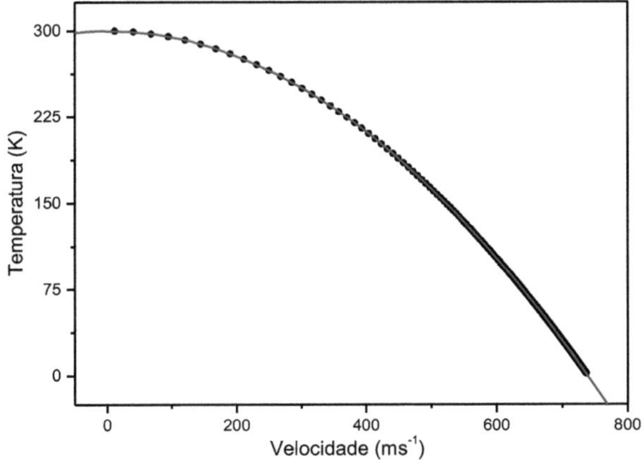

Figure 12 - Simulation of temperature as a function of velocity in an SMB for diatomic molecules.

By applying the least squares method to the graph of temperature as a function of speed, a direct relationship can be established between speed and temperature. Temperature as a function of speed is given by:

$$T = 300 - 5{,}5x10^{-4}(u)^2 - 1{,}07x10^{-10}u \qquad (31)$$

In SMB, the dynamic cooling process associated with supersonic expansion reduces the local speed of sound of the gas, substantially increasing the Mach number. At the same time, the random movement of the particles is converted into movement in the direction of the flow, resulting in an increase in the average flow velocity.

2   Objectives

The aim of this work is to determine the velocities of flows produced inside a vacuum expansion chamber. The chamber characterisation database will be used to validate other optical techniques used in velocity measurements. In addition to the classic method based on the use of a fast ionisation detector, velocity values were determined using the non-intrusive Schlieren optical technique. Finally, in order to investigate flows in which the conventional optical schlieren method is not applicable, a modification to the method was proposed, based on the introduction of absorbing molecules into the flow.

3   Experimental

All the experimental work was carried out at the Institute for Advanced Studies of the Department of Aerospace Science and Technology, as part of the research activities linked to the HIPERVEL project (Hypersonic Flow Velocity Measurements). Annex A shows photographs of the equipment used in the experiments, as well as the respective experimental setups.

3.1   Vacuum System

The pumping system used to produce a vacuum in the expansion chamber consists of an Edwards mechanical rotary vane pump, model E2M2, with which it was possible to reach a pressure of *ca.* 10^{-3} mbar in the expansion chamber and an EMT - Tecnologia Eletromecânica Ltda diffuser pump, with a pumping speed of 1800 litres/s, capable of reaching a pressure of *ca.* 10^{-6} mbar. The pressure measurement system consists of two detectors: a digital Piezo/Pirani vacuum gauge from Oerlikon Leybold Vacuum, model thermovac tm 101, with a measurement range of 1200 to 5×10^{-4} mbar and a vacuum gauge with a *full range* Pirani/cold method sensor from Pfeiffer, model PKR 251, with a measurement range of 1000 to 5×10^{-9} mbar, coupled to a control unit from Pfeiffer, model TPG 261.

Each detector has a characteristic wide pressure measurement band, outside of which the system does not provide accurate measurements. In the digital meter, the optimum operating range is between 1000 and 10^{-3} mbar, while for the *full range* meter the optimum *range* is 10^{-2} to 10^{-8} mbar.

Figure 13 shows the experimental procedure for switching the expansion chamber vacuum system on and off. It is important to follow the steps in the procedure to avoid contamination of the expansion chamber by oil from the diffuser pump.

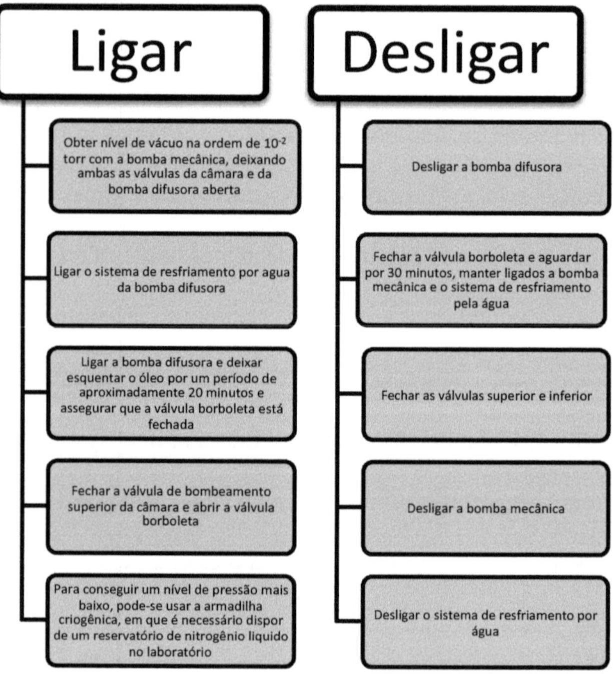

Figure 13 - **Procedures for switching the vacuum system on and off.**

To achieve lower pressure values, liquid nitrogen can be added to the chamber's cryogenic trap. Liquid nitrogen can reduce the pressure in the chamber by up to an order of magnitude, reaching 10^{-7} mbar. At the same time, its use prevents oil from contaminating the chamber.

3.2   Pulsed valve

The pulsed valve system consists of a pulsed valve and a Lasertechnics model 203B controller. To generate the opening and control the number of pulses per second, a wave function generator connected to the valve controller is required. The generator used was a Goldstar model FG2002C. Table 3 shows the operating values for the pulsed valve controller [56].

Table 3 - Operating system values for the pulsed valve driver.

Operating system values	
Controller voltage	110 V
Minimum wrist width	150 µs
Current peak	1 A
Dissipation energy	less than 10 mW
Repetition rate	0-500 pulses per second
Maximum operating temperature	393 K
Maximum voltage of a pulse	180 V
Operating pressure	0 to 10 atm

The valve controller has three operating modes: *Follow, Delay* and *Oscillate.* In *Follow* mode, the signal going to the valve is the same as the signal from the wave function generator. In *Oscillate* mode, the pulse profile varies as a function of time. In *Delay* mode you can control the delay of the valve signal in relation to the function generator signal. In this mode it is possible to obtain the smallest temporal width of the valve pulse, which is 150 μs.

3.3 Experimental arrangement for velocimetry by Fast Ionisation Detector (FIG)

The experimental setup (Figure 14) consisted of a stainless steel expansion chamber, vacuum system, pulse valve system and the flow detection system. The detection system consisted of the FIG and a Beam Dynamics controller. In previous work using the expansion chamber and FIG used in this study, speeds of up to *ca.* 720 ms were measured [-1 54].

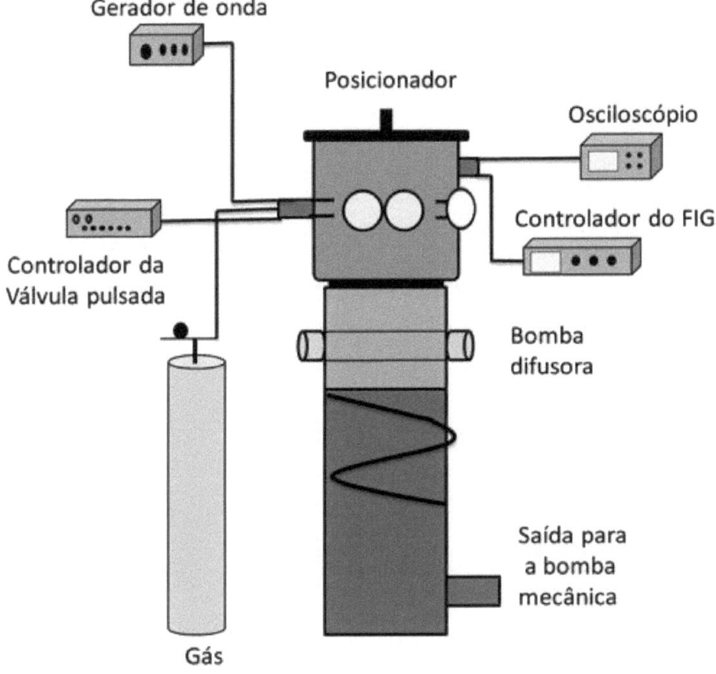

Figure 14 - Experimental setup for velocimetry using FIG.

Figure 15 shows the Rapid Ionisation Detector. The hot filament produces electrons that are accelerated by a potential difference of 160 V towards the grid. As the gas passes through this region, it collides with the electrons, ionising the molecules. The ions formed there travel towards the potential grid. In the space between the grids is the collector, which has a potential of 0 V. The detector's operating pressure range is between 10^{-4} and 10^{-8} mbar [57].

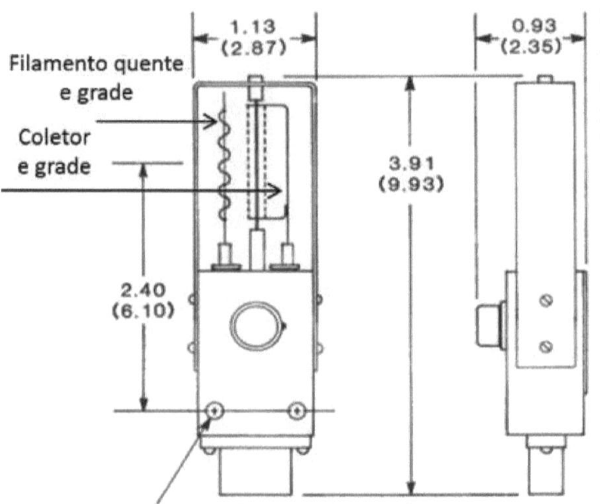

Figure 15 - Rapid Ionisation Detector. [57]

From the known distance between the valve and the detector, and the time of flight of the molecules, obtained by measuring the time interval from the start of the controller pulse and the detection of the flow by the FIG, the velocity, temperature and Mach number of the jets were calculated. The flow was produced using pure oxygen (99%) as the gas.

To characterise the flows produced in the expansion chamber, the chamber pressure was kept constant at 10^{-6} mbar. and the stagnation pressure and valve opening time were varied. The stagnation pressure of the valve's operating gas was varied by 2, 4, 6 and 8 bar. The valve opening time was varied by 100, 150, 250 and 300 µs. The pulse frequency was kept constant at 10 Hz. The voltage applied to open the valve was kept constant at 160 V.

3.4   Experimental setup for Schlieren Velocimetry

Velocimetry based on images obtained using the Total Field Schlieren technique made it possible to characterise flow velocity in a non-intrusive way. Figure 16 shows the optical arrangement used in the experiment, where L_1 , L_2 and L_3 are lenses, M_1 and M_2 are mirrors and K is the knife.

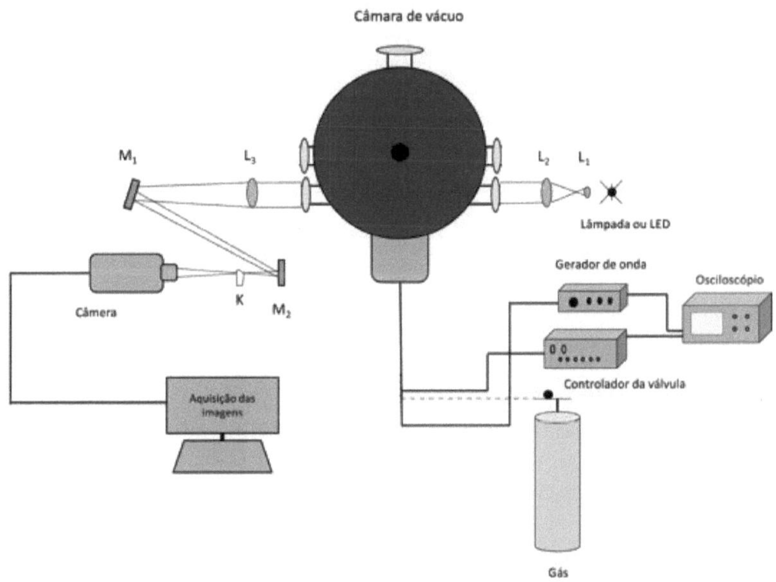

Figure 16 - Experimental setup for schlieren velocimetry.

Various incandescent lamps and LEDs were tested as a light source in order to find the one with the best contrast. To produce the parallel light beam that passes through the test area, a beam expander was used, corresponding to lenses L1 and L2. Lens L3 was used to focus the parallel light beam. Different lenses were tested, with focal lengths of 20, 30, 50, 100 and 200 mm, in order to find one that best presented contrast in the image and at the same time was compatible with the space of the optical table available in the laboratory. A spatial filter (knife) was positioned at the focal point of the lens to block light rays that had been deflected. A PCO *Sensitivity* CCD camera and a Phantom model MRLC110 high-speed camera were used for image acquisition.

The flow velocity was determined by a simple procedure, based on the displacement of the front of the flow recorded in consecutive images and the time between them, given by the camera's capture rate. Given the large volume of images obtained, this procedure was automated using the camera's control programme. In order for the programme to provide the velocity data, a calibration was required between the number of pixels in the image and a reference distance. A ruler was used for this, Figure 17. The gas used in the flow was pure trichlorofluoromethane (CCl_3 F).

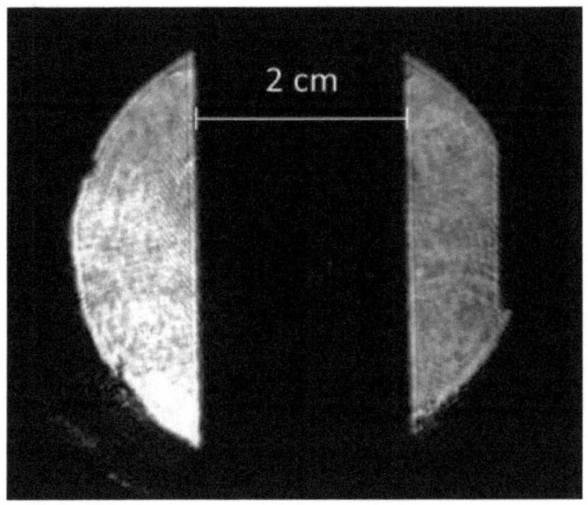

Figure 17 - Calibration to calculate the relationship between number of pixels and distance, the ruler was placed in front of the valve.

3.5   Experimental setup for SCAM velocimetry

Figure 18 shows the experimental setup for Schlieren Total Field Velocimetry Combined with Molecular Absorption. The optical setup used in the experiment was similar to that used in schlieren and image acquisition was carried out by a Phantom model v411 fast camera.

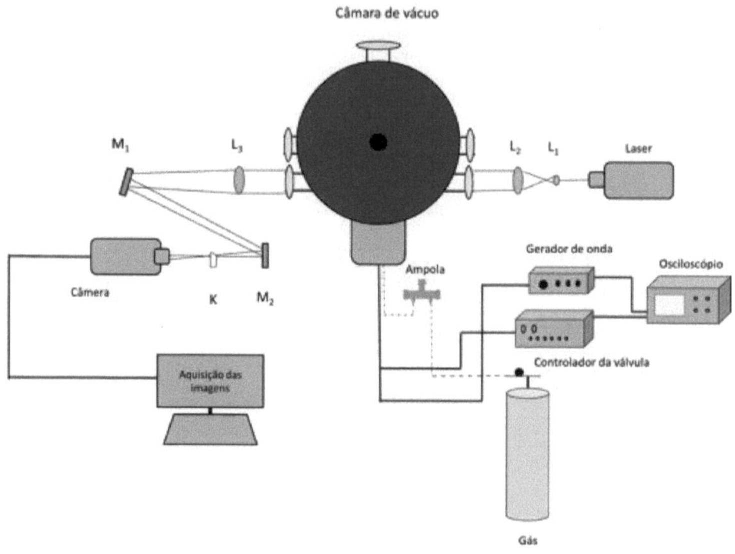

Figure 18 - Experimental setup for schlieren velocimetry combined with molecular absorption.

Firstly, the effect of wavelength on image contrast was investigated. Two continuous lasers were used as the light source: a helium neon laser (HeNe), emitting at a wavelength of 632.99, and a neodymium laser doped with yttrium orthovanadate (Nd:YVO$_4$), from the Coherent brand, model V8, emitting at a wavelength of 532.15 nm. The flow gas was pure nitrous oxide (N$_2$ O).

In the experiments involving absorption, the flow was seeded with I molecules$_2$. The seeding process was carried out by sublimating solid iodine. Figure 19 shows an ampoule heated by a blower, where SF gas$_6$ at high pressure passes through the tube, dragging the sublimated iodine molecules.

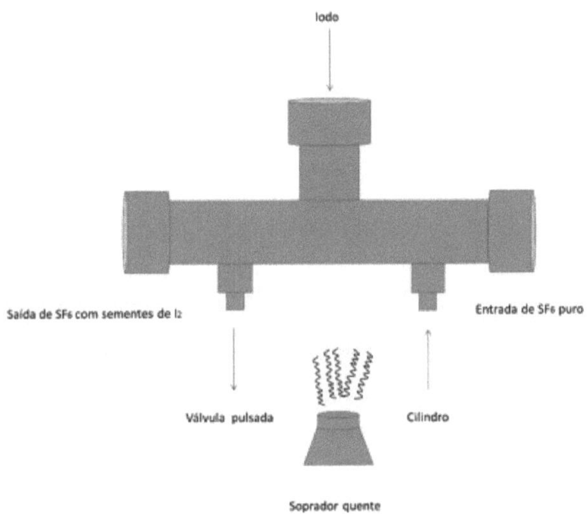

Figure 19 - Ampoule used for seeding I$_2$ in SF runoff$_6$.

4 Results and Discussions

The first stage of assembly consisted of reactivating the expansion chamber vacuum system, which had been deactivated for around 15 years. The burnt-out mechanical pump motor and the diffuser pump oil were replaced. All the vacuum chamber rings were then replaced and the optical windows made in the IEAv optics workshop.

The second stage consisted of assembling the pulsed valve system, whose electrical circuit was damaged. The third stage consisted of testing and operating the system FIG. Before this, however, it was necessary to replace several of its electronic components, which were burnt out. At this stage it was necessary to study their operation in detail, since production of these devices was discontinued years ago. After replacing the damaged part of the detector, the entire Velocimetry experimental system using a Rapid Ionisation Detector was operational.

4.1 Velocimetry using a Rapid Ionisation Detector

Initially, the flow produced by the pulsed valve in the supersonic regime was characterised in order to compare it with other work carried out at the Institute for Advanced Studies with SMBs [48, 54, , ,585960]. The aim of this study was to characterise the supersonic flow produced in an expansion chamber at different stagnation pressures and valve opening times.

The first measurements using the system were aimed at determining the valve's pulse profiles. Figure 20 shows a typical result of the pulse profile obtained by FIG, using oxygen gas as the operating gas. The valve opening time was 150 µs, with a stagnation pressure of 6 bar. The detector was positioned at a distance of 167 mm from the valve.

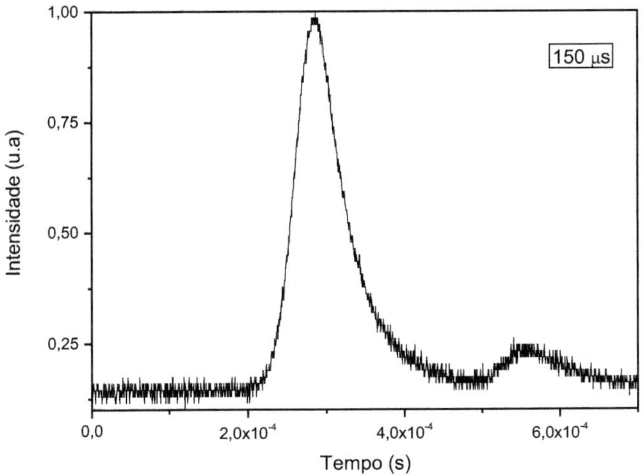

Figure 20 -Typical pulse profile recorded by FIG. Valve opening time, 150 µs; stagnation pressure, 6 bar; distance of detector from valve outlet, 167 mm.

According to Figure 20, the measured temporal width of the pulse is 150 µs. The signal obtained by the detector is a convolution between the velocity distribution function (VDF) and a modulation function. The experimental VFD corresponds to the number of molecules reaching the detector at a given point as a function of time, while the modulation function corresponds to the sum of the valve and detector opening signals. Studies involving FIG signal profiles obtained from pulses from a valve of the same model used in this work have shown that the signal profile is a convoluted Maxwellian velocity distribution function with a trapezoidal profile modulation function [54].

To measure the time of flight of the molecules in the flow, a programme was developed to read the data stored on the oscilloscope's memory card. The system's delay time, corresponding to the time between the controller sending the signal and the valve opening, was obtained from the literature, with a value of 60 µs [61]. Figure 21 shows the signals obtained by the oscilloscope. The upper curve corresponds to the signal sent by the controller to the pulsed valve and the lower curve corresponds to the signal recorded by the FIG. The detector was positioned at a distance of 167 mm from the valve outlet. The stagnation pressure was 6 bar and the valve opening time was 150 µs. The time of flight corresponds to the time interval of 265 µs between the start of the valve opening signal and the start of the signal obtained by the FIG.

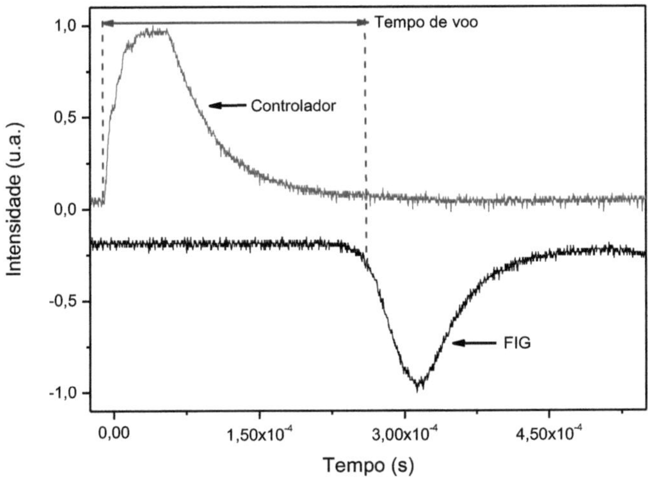

Figure 21 - Signals obtained by the valve controller and recorded by FIG. Valve opening time, 150 µs; stagnation pressure, 6 bar; distance of the detector from the valve outlet, 167 mm.

From the time of flight it was possible to calculate the velocity, temperature and Mach number of the flow. For a given value of stagnation pressure and valve opening time, the time of flight was determined experimentally at sixteen positions along the flow axis, moving the valve longitudinally.

Table 4 shows the results of the average speed, in 16 different positions from 6 to 167 mm. For a valve opening time of 150 µs. The stagnation pressure was varied by 2, 4, 6 and 8 bar. The error associated with the speed was estimated from the standard deviation of at least six measurements. Considering all the sets investigated, the highest standard deviation was *ca.* 3 %. This value was adopted as the relative error of the speed values. In the blank gaps it was not possible to obtain the velocity due to problems with the FIG signal. The other velocity values for the other valve opening times at 100, 200, 250 and 300 µs are in Appendix B.

Table 4 - Average velocity measurement of O molecules$_2$ as a function of distance and stagnation pressure, with valve opening time constant at 150 µs.

Pressure (bar)	2	4	6	8
x (mm)	Speed (ms)$^{-1}$			
6	58±2	57±2	64±2	62±2
16	121±4	115±3		127±4
26	157±5	179±5	211±6	
36	243±7		255±8	265±8
46	287±9	300±9	292±9	299±9
56	380±11	412±12		309±9
67	400±12	406±12	401±12	396±12
77	481±14		459±14	465±14
87	584±18	522±16	530±16	481±14
97	626±19	563±17	551±17	537±16
117	617±19	568±17	567±17	559±17
127	634±19	546±16	599±18	567±17
137	670±20	617±19	587±18	572±17
147	589±18	701±21	668±20	633±19
157		700±21	672±20	668±20
167	720±22	687±21	632±20	690±21

The results in Table 4 indicate that the velocity profile does not vary as a function of the stagnation pressure studied. Figure 22 shows the calculated velocity profile for a stagnation pressure of 8 bar with opening times of 100, 150 and 300 µs.

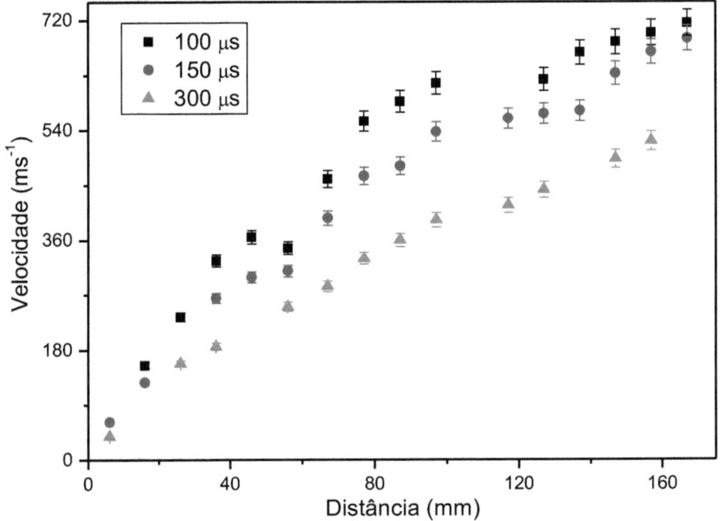

Figure 22 - Average velocity of O molecules$_2$ for pulses of 100, 150, 300 µs and stagnation pressure 8 bar.

The graph in Figure 22 shows that the velocity within the range of positions investigated increases with the distance travelled. As the flow propagates, collisions occur that transfer the energy stored in the molecule's internal modes to the translational mode. This process causes the rotational and vibrational temperatures of the molecules to cool [37], (Figure 12).

Comparing the experimental results with those obtained theoretically, it can be seen that there is a discrepancy in the velocity profile. Several points can help interpret the difference between the two.

(i)   Firstly, a *skimmer* was not used in the experiments. This device is a bulkhead that acts as a spatial filter for the flow. Its function is to collimate and select a region of the flow. For SMB applications, the central region of the flow is normally selected, where the velocity distribution is smaller [38]. In addition, the centre of the flow has a higher velocity than the edges. In a measurement using a *skimmer*, the flow is expected to reach the free molecular regime condition in a region close to the valve outlet, which is close to the theoretical prediction.

(ii)   The Characteristics Method considers a one-dimensional expansion of an ideal gas. However, in a real gas, you have to consider that the flow is a three-dimensional expansion. For a real gas, for example, viscosity effects can cause a decrease in velocity.

(iii)   The shape of the valve's gas outlet nozzle contributes to the orifice discharge factor, which is the valve's flow parameter [,6263]. Consequently, the density of the flow can vary. For a flow in which the density is higher, it is expected that the length of the collision region of the molecules will be greater. This means that the flow velocity is lower.

Table 5 shows the Mach number and flow temperature calculated in 16 different positions with a valve opening time of 150 µs and stagnation pressures of 2 and 8 bar. The temperature is calculated from equation (31), which is derived from the least squares method applied to the graph of temperature as a function of speed. The Mach number was calculated from equation (23) using the velocity values in Table 4 and the temperatures calculated using the previous procedure. The error in the calculations was determined using the error propagation method.

Table 5 - Mach number and temperature calculated at stagnation pressures of 2 and 8 bar, with pulse duration constant at 150 μs.

Pressure (bar)	2	2	8	8
x (mm)	M	T (K)	M	T (K)
6	0,18±0,01	298±9	0,19±0,01	298±9
16	0,37±0,01	292±9	0,39±0,01	291±9
26	0,49±0,01	286±9		
36	0,78±0,02	268±8	0,86±0,03	261±8
46	0,94±0,03	255±8	0,99±0,03	251±8
56	1,34±0,04	220±7	1,03±0,03	247±7
67	1,44±0,04	212±6	1,42±0,04	214±6
77	1,92±0,06	173±5	1,81±0,05	181±5
87	2,89±0,09	112±3	1,92±0,06	173±5
97	3,57±0,11	85±3	2,37±0,07	141±4
117	3,4±0,10	91±3	2,59±0,08	128±4
127	3,74±0,11	79±2	2,68±0,08	123±4
137	4,83±0,14	53±2	2,74±0,08	120±4
147	2,96±0,09	109±3	3,71±0,11	80±2
157			4,72±0,14	55±2
167	9,74±0,29	15±1	5,86±0,18	38±1

Figure 23 shows the Mach number as a function of distance from the orifice, with a stagnation pressure of 8 bar and valve opening times of

100, 150 and 300 μs. The graph shows that the Mach number increases more sharply from a distance of 127 mm. The maximum value calculated for the Mach number is 10.7 for a valve opening time of 100 μs and a stagnation pressure of 2 bar, see appendix B. Comparing the results obtained with the theoretical ones, it can be seen that the calculated Mach number value is approximately one third of the theoretical value. The Mach number is calculated from the speed in the graph in Figure 22. Therefore, the discrepancy between the experimental and theoretical values can be justified by the same arguments used to calculate the speed.

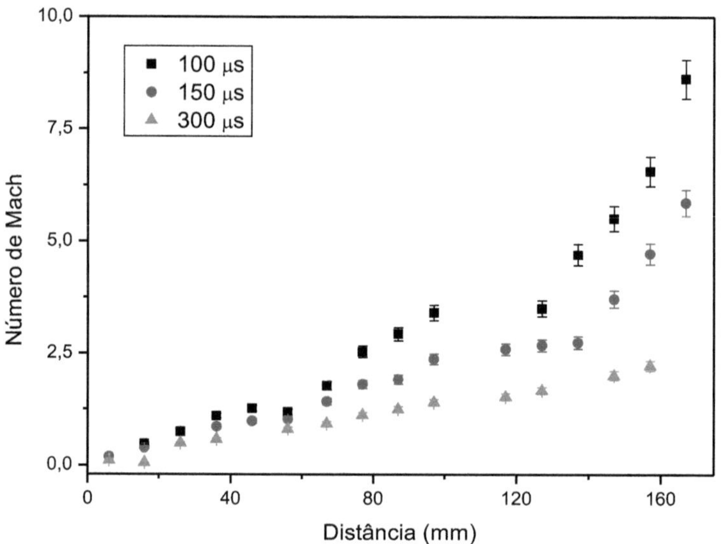

Figure 23 - Mach number as a function of distance from the valve at 8 bar pressure.

Figure 24 shows the temperature as a function of distance from the orifice, with a stagnation pressure of 8 bar and valve opening times of 100,

150 and 300 µs. Contrary to what was theoretically predicted (Figure 10), the temperature shows a decay that is close to a linear behaviour. This discrepancy is associated with considerations about the physical-dynamic nature of the jets investigated.

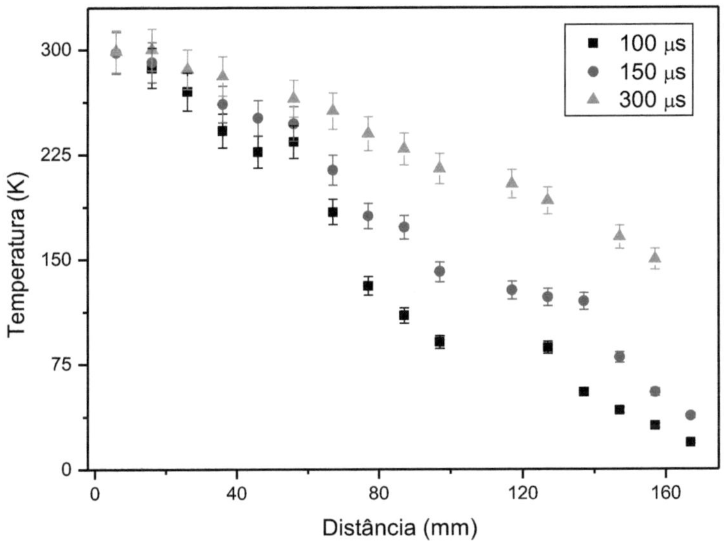

Figure 24 - Temperature as a function of distance from the valve at 8 bar pressure.

Analysing the graphs in Figures 22, 23 and 24, and the results presented in the annex, it can be seen that for the valve opening time of 100 µs, the values of u, M and T are closer to the respective theoretical values. Consequently, the 100 µs pulses are closer to the free molecular flow regime condition. This result reinforces the hypothesis that, for higher flow densities, the extent of the collision region is greater.

The results of the flow study using the Fast Ionisation Detector showed that it was possible to obtain supersonic flows in continuous and transition regimes. However, at the distances investigated, it was not possible to distinguish a region of constant velocity, as predicted by the theoretical model. This indicates that collision processes are present throughout the entire range of distances travelled by the flow and therefore the flow has not reached the free molecular regime.

4.2   Schlieren Velocimetry

Initially, the schlieren optical technique was applied to the flow of oxygen molecules produced under the same conditions as above. However, even after exhaustive attempts, it was not possible to visualise the flow.

The first attempt to change the experiments in order to visualise the flow was with regard to the optical arrangement. According to equation (12), sensitivity is proportional to the focal length of the lens used to project the object of the test section. Lenses of 25, 50, 75 and 100 cm were investigated. However, no lens was able to make the flow visible. The limit value for the focal length of the lenses was limited by the size of the optical table used in the experiments.

Alternatively, the jets were produced under higher pressure values in the vacuum chamber. To achieve these conditions, the diffuser pump was switched off, and at the same time a millimetre exhaust valve was attached to the chamber, responsible for adjusting the chamber pressure values. According to equation (11) for the contrast of the schlieren images and equations (3) and (6) for the refraction of the light rays, for denser media the variation in the refractive index is more intense. Consequently, schlieren images are more easily visualised at higher pressures. The experiments showed that, using a lens with a focal length of 1 metre, from a pressure of 500 mbar, it was possible to visualise the flow of oxygen gas.

In the search for greater image sharpness, various light sources were investigated. Various incandescent and LED lamps were tested, and the best result was obtained with an LED taken from a commercial optical mouse, coupled to a beam expander (Figure 16).

As a result of the increase in chamber pressure, there was a change in the flow velocities produced by the valve. Since the velocity is a function of the difference between the stagnation pressure and the chamber pressure, at lower chamber pressures the flow velocity is higher, but the contrast is lower. To reduce this effect, it was decided to change the operating gas so that the pressure threshold for visualising the flow was lowered. The following gases were tested: helium (He), oxygen (O_2

), nitrogen (N_2), nitrous oxide ($N_2 O$), sulphur hexafluoride (SF_6) and trichlorofluoromethane ($CCl_3 F$).

According to equation (2), the contrast in the schlieren image is directly proportional to the refractivity values (*n-1*). The refractivity values multiplied by a factor of one thousand are shown in Table 6. The values were obtained from the refractive index of gases available in the literature [64].

Table 6 - Refractoriness of gases multiplied by a factor of 1000.

Gas	He	O_2	N_2	$N O_2$	SF_6	$CCl F_3$
(n-1)*1000	0,3197	2,712	2,732	4,653	7,83	12,1

The dependence of contrast on refractoriness was confirmed by the experimental results. It was therefore decided to use trichlorofluoromethane, which showed the greatest contrast in the schlieren images. Initially, the images were acquired by a PCO *Sensitivity* CCD camera. Figure 25 shows the image of the trichlorofluoromethane flow obtained at a pressure of 300 mbar and a stagnation pressure of 2 bar.

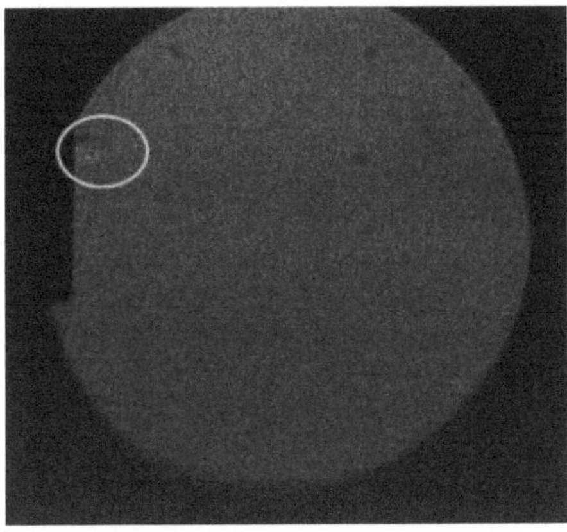

Figure 25 - Schlieren image of CCl flow₃ F, using the PCO camera.

In order to determine the speed between two points using the schlieren technique, you need at least two images in which you can fix the image capture time and the interval between the two shots. Due to its technical characteristics, the PCO camera allows only two images to be taken per pulse. Once the time of the start of the expansion has been determined, the aim would be to insert different delay values in the acquisition of the second image, in different pulses, in order to map the speed of the flow at different points. However, the PCO camera presented two defects during the experiments. The first problem was synchronisation with the valve controller. The second, of a practical nature, concerned the saturation of the images obtained. When the camera was exposed to an amount of light above its saturation threshold, a common occurrence in the search for greater sensitivity, the image showed permanent blurring. The images taken consecutively maintained the same blur patterns, which made it impossible to accurately determine the distance travelled by the flow.

To get round this problem, the PCO camera was replaced by a high-speed Phantom camera. With a maximum rate of 400,000 frames per second and a minimum exposure time of 2 µs, the camera made it possible to study the entire expansion of a single pulse.

Figure 26 shows the schlieren images obtained by the high-speed camera at a rate of 36,000 frames per second with an exposure time of 3.42 µs. The flow was produced from a stagnation pressure of 3 bar and with the chamber operating at ambient pressure and a valve opening time of 2 ms. When determining the velocity, the expansion of the flow at the valve outlet was taken as the initial time.

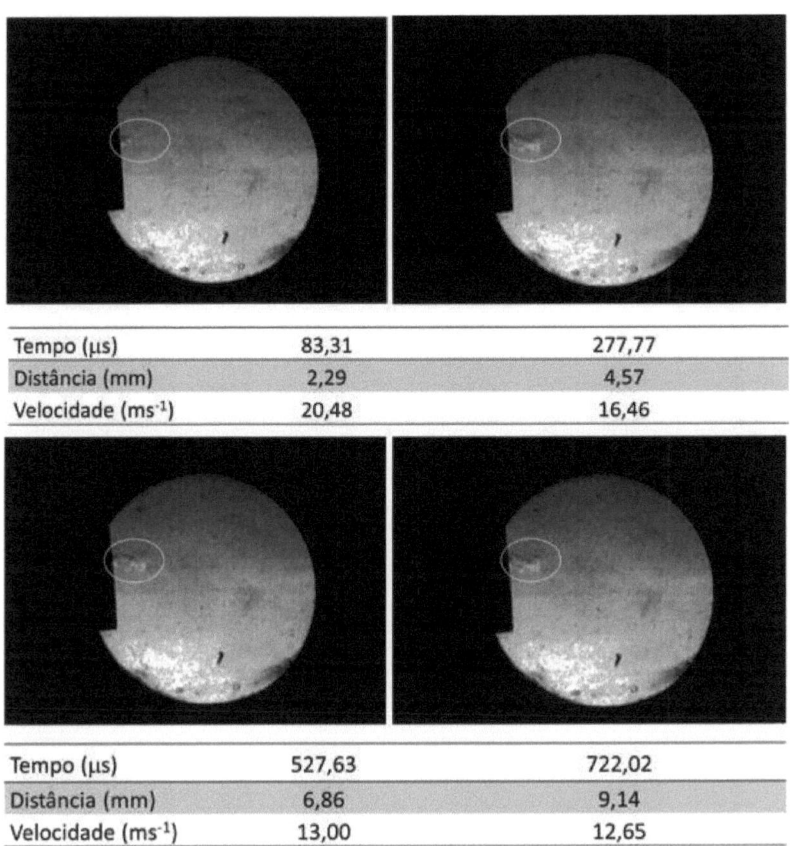

Tempo (µs)	83,31	277,77
Distância (mm)	2,29	4,57
Velocidade (ms⁻¹)	20,48	16,46

Tempo (µs)	527,63	722,02
Distância (mm)	6,86	9,14
Velocidade (ms⁻¹)	13,00	12,65

Figure 26 - Schlieren flow visualised at different times.

The chamber window through which the flow is viewed has a diameter of 2 inches. The maximum distance for viewing the flow depends on the parameters of stagnation pressure, chamber pressure and valve opening time. Figures 27 and 28 show the average velocity of the trichlorofluoromethane molecules. In Figure 27 the stagnation pressure was 2 bar and the chamber pressure was 1 bar. In Figure 28 the stagnation pressure was 3 bar and the chamber pressure was 1 bar. The other velocity

calculations can be found in Appendix C.

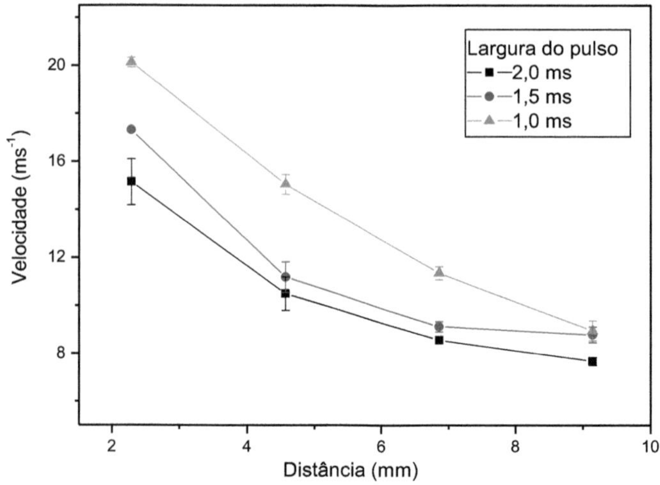

Figure 27 - Average velocity of trichlorofluoromethane molecules at 2 bar stagnation

pressure and 1 bar chamber pressure .

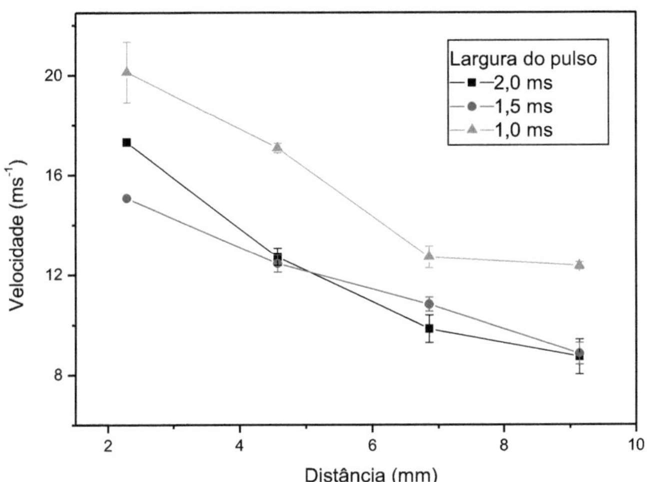

Figure 28 - Average velocity of trichlorofluoromethane molecules at 3 bar stagnation pressure and 1 bar chamber pressure.

The average speed of the molecules decreases throughout the expansion. In this process, the molecules collide with the gas at the bottom of the chamber, which acts as a bulkhead for the molecules, thus reducing the flow velocity. The maximum speed observed is close to the valve outlet region, *ca.* 20 ms^{-1} . The maximum distance at which the flow could be visualised was 9 mm. After this value, the density of the flow and, consequently, its contrast, makes it impossible to precisely identify the flow front.

Finally, the highest velocity values were recorded for the shortest opening times, a similar behaviour observed in flows produced at low pressure (detection with FIG). The error of the measurements was determined from the standard deviation of 6 velocity measurements of the same pulse.

Figures 29 and 30 show the average velocity of the trichlorofluoromethane molecules. In Figure 29, the stagnation pressure was 3 bar and the chamber pressure was 300 mbar. In Figure 30, the stagnation pressure was 3 bar and the chamber pressure was 250 mbar.

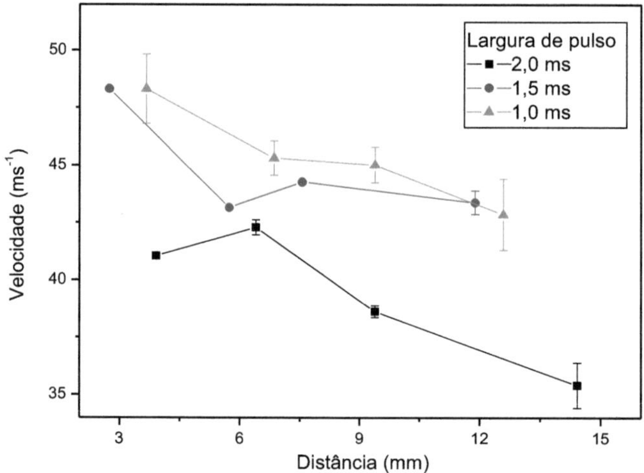

Figure 29 - Average velocity of trichlorofluoromethane molecules at stagnation pressure 3 bar and chamber pressure 300 mbar .

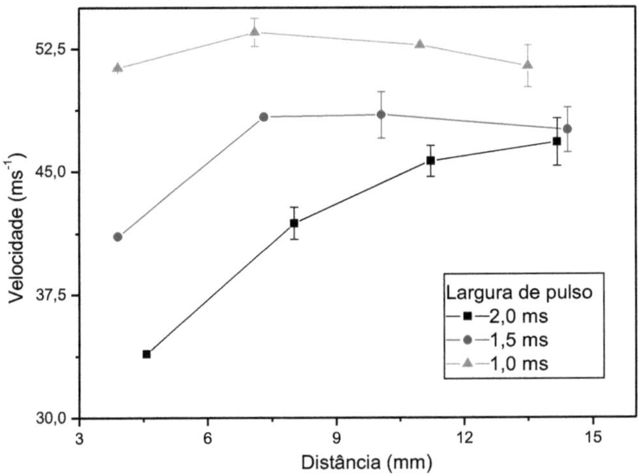

Figure 30 - Average velocity of trichlorofluoromethane molecules at stagnation pressure of 3 bar and chamber pressure of 250 mbar.

Under these pressure conditions, as shown in Figures 29 and 30, there is a competition between deceleration caused by the collision with the background gases and acceleration due to the expansion process in which there is a transfer of energy from the internal modes to the translational mode. In Figure 30, the lower pressure (250 mbar) means that the flow acceleration mechanisms are greater than the deceleration mechanisms. Consequently, the velocity curve as a function of distance from the valve has a positive slope.

Table 7 shows the Mach number for the 2 ms pulse, calculated at different stagnation pressures p_{est} and internal chamber pressure $p_{câm}$. The other calculations are in Appendix C. The temperature was determined using equation (31). For the speed range investigated, the temperature varied little, with the minimum value being 298 K and the maximum 300 K.

Table 7 - Mach number.

$p_{câm}$	1 bar	1 bar		300 mbar		250 mbar
p_{est}	3 bar	2 bar		3 bar		3 bar
x (mm)	Mach	Mach	x (mm)	Mach	x (mm)	Mach
2,286	0,137±0,008	0,046±0,003	3,658	0,147±0,005	3,901	0,125±0,000
4,572	0,116±0,001	0,032±0,002	6,858	0,137±0,002	7,315	0,147±0,000
6,858	0,086±0,003	0,026±0,001	9,388	0,137±0,002	10,058	0,147±0,004
9,144	0,084±0,001	0,023±0,001	12,588	0,130±0,005	14,417	0,144±0,004

In the speeds investigated using the schlieren optical technique, the maximum value determined was 65.5 ms^{-1} at a temperature of 298 K. Consequently, the flow investigated is subsonic, and the maximum value of the Mach number was 0.147 for a stagnation pressure of 250 mbar.

Varying the chamber's internal pressure using a micrometric valve proved to be an efficient mechanism for controlling flow velocity. However, varying the values of the stagnation pressures investigated had little influence on the flow velocity.

Below a pressure of 250 mbar it was not possible to accurately determine the value of the distance travelled by the flow. Figure 31 shows the velocity values for a stagnation pressure of 200 mbar. This pressure is the threshold for visualising the flow, and the difficulty of identifying the flow front can be seen in the high standard deviation associated with the velocity calculation.

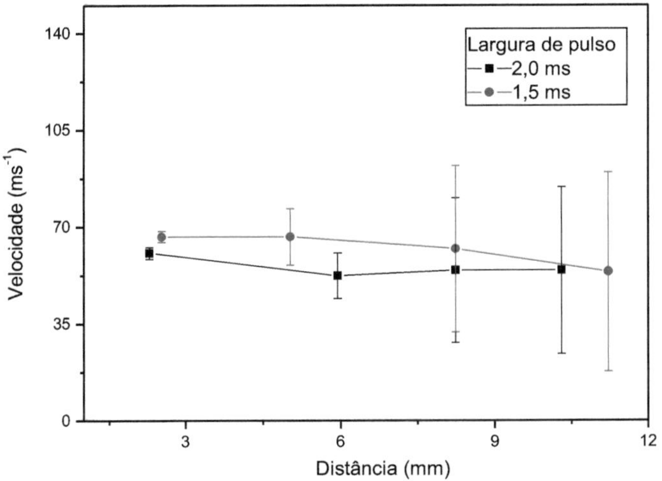

Figure 31 - Average velocity of trichlorofluoromethane molecules at stagnation pressure 3 bar and chamber pressure 200 mbar

4.3   Schlieren Velocimetry Combined with Molecular Absorption

In order to investigate flows produced at lower pressure values than conventional schlieren, a modification of the method was proposed. Initially, the contribution of the wavelength of the light source to image contrast was studied. Subsequently, simulations were carried out on the iodine spectrum to investigate temperature effects on the wavelength absorption line of the light source. The third stage consisted of investigating the flows using the optical SCAM method.

4.3.1 Contrast comparison using two lasers

The first stage of the experiment consisted of investigating the contrast in the image of a flow using two lasers as a light source, at wavelengths of 532.15 nm (corresponding to the colour green) and 632.99 nm (corresponding to the colour red). A flow consisting of pure nitrous oxide, expanded in the chamber at atmospheric pressure, was investigated. It was chosen because of its refractivity value and availability for use.

To compare the contrast using different lasers, the knife was positioned so as to guarantee the best contrast obtained by the arrangement. The power of both lasers was 10 mW. Figure 32 shows the schlieren visualisation of the flow, using (a) the green laser and (b) the red laser as the light source. In both cases, the images were acquired at times of 500, 1225 and 2100 μs from the start of the flow. Additional images are shown in Appendix D to illustrate the flow expansion process using the green laser as the light source.

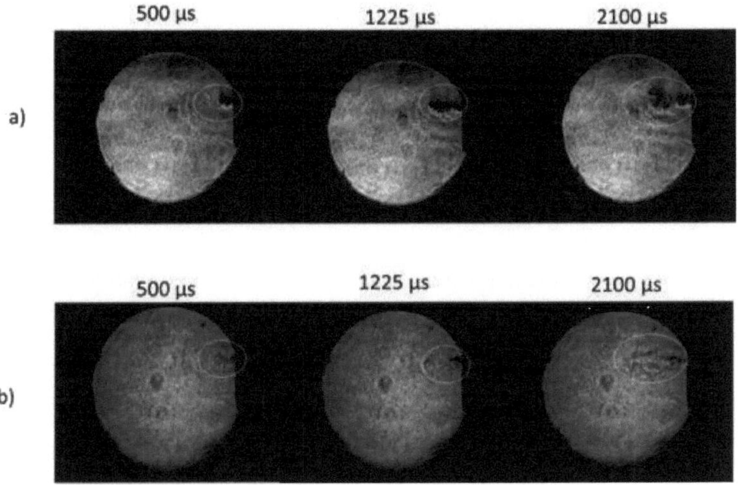

Figure 32 - Comparison between the schlieren images of the flow, (a) laser at 532.15 nm, (b) laser at 632.99 nm.

The contrast obtained with the green laser is higher when compared to the images obtained with the red laser. This result is in line with equation (2), in which the Gladstone-Dale coefficient k decreases slightly with increasing light wavelength λ. For larger values of k, the refraction is greater and consequently the contrast is greater.

As a result, the green laser was chosen to be used as the light source in the schlieren experiments. In parallel, a search of the literature had shown the use of the I_2 molecule as a spectral filter in Filtered Rayleigh Scattering Velocimetry. In this technique, I_2 is used as a filter to block the spectral broadening contribution of Mie scattering, allowing only Rayleigh scattering to pass through [65]. This gave rise to the idea of using it as an absorber species in the flows investigated, thus giving rise to the SCAM method.

4.3.2 Absorption spectra of I_2

Due to its large absorption band in the visible, the iodine molecule can be used as an ultrafine filter at various wavelengths, which enables it to be used in various laser applications [66]. The electronic spectrum of the iodine molecule shows absorption in the visible region, with progressions of bands involving vibrational states. Due to its high mass, the rotational structure is not resolved, but is responsible for the sawtooth shape of the vibrational components of the electronic spectrum. The electronic spectrum shows an intense absorption maximum in the green region [67] and strong absorption in the 532.15 nm laser wavelength region [,6869].

In order to study the region of laser absorption in iodine molecules, the spectrum of this region was simulated using the IodineSpec 5 programme. The programme is capable of simulating spectra with rotational and hyperfine structures in the 514 to 892 nm range [70]. The programme was developed from the database of Gerstenkorn and Luc's work [,7172]. Figure 33 shows the simulation of the rotational absorption spectrum of iodine in the wavelength range 532.025 to 532.305 nm. The simulation was carried out at a temperature of 300 K. The linewidth of the spectra simulated from the programme is the sum of the Doppler effect and the natural one.

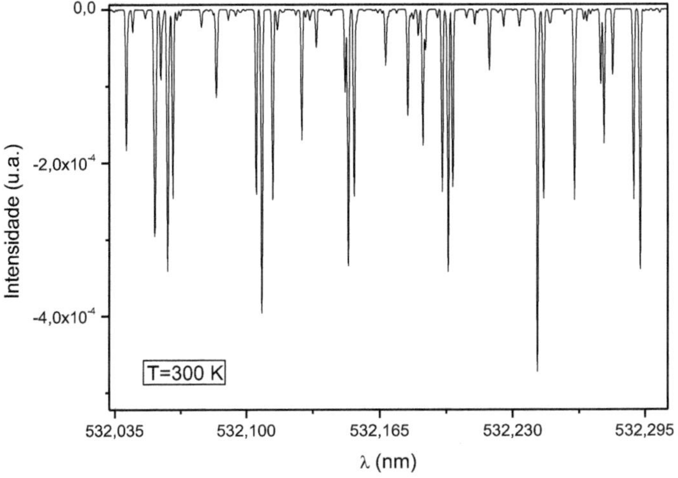

Figure 33 - Iodine absorption spectrum in the range 532.0250 to 532.3050 nm.

The region of iodine excited by the Nd:YVO4 laser was at a wavelength of 532.15 nm. This region corresponds to 3 absorption peaks, Figure 34. The absorption doublet in this region is quite intense, as it has a strong absorption coefficient which reduces the transmittance of the light at this wavelength.

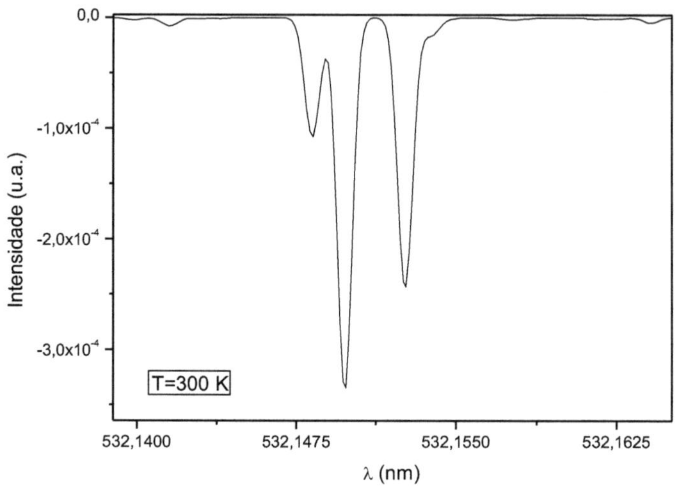

Figure 34 - Iodine absorption peaks in the region of 532.15 nm.

In the experiments using the fast ionisation detector, temperatures of around 20 K were calculated for the flows produced under a vacuum of 10^{-6} mbar. With the schlieren technique, flows with a temperature of 300 K were studied. To study the variation of the absorption coefficient at different temperatures, the intensities of the doublet peaks from 10 to 300 K were simulated. Figure 35 shows the relationship between peak absorption intensity and temperature, where the intensity was normalised to 300 K.

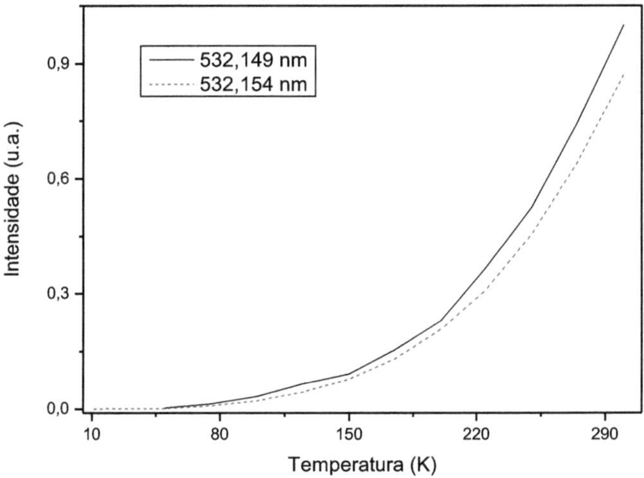

Figure 35 - Absorption intensity as a function of temperature for iodine molecules.

As can be seen, the absorption intensity decreases exponentially as the temperature decreases. Consequently, in flows close to the free molecular regime, absorption is close to zero. Therefore, contrast enhancement using the SCAM method with iodine seeds is more effective in situations where the flow is in the continuous regime, i.e. at higher temperatures.

4.3.3 Determination of SF flow velocity$_6$ with I seeds$_2$

The working gas used was SF_6 instead of $N_2 O$, as it is more refractory. An ampoule containing solid iodine heated by a hot air blower was used to seed the flow. As it passes through the ampoule, the pressurised hexafluoride drags the iodine molecules, which are carried through a flexible polyethylene tube to the valve. The flow studied was produced at a stagnation pressure of 5 bar, with a valve opening time of 2 ms and chamber pressures of 30, 20 and 15 mbar.

Figure 36 shows the schlieren image of the flow, at an internal camera pressure of 30 mbar. The image was taken 500 µs after the start of the expansion. In Figure 36 (a) the flow consisted only of pure SF_6, while in Figure 36 (b) the flow was seeded with iodine molecules.

a)SF6 puro
b)SF6 com sementes de I2

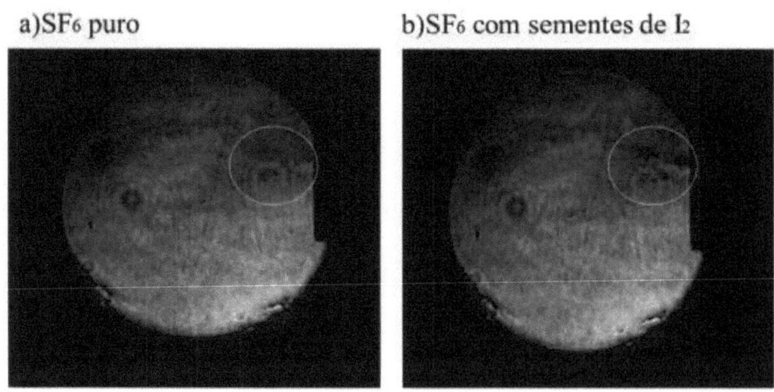

Figure 36 - Schlieren image of the flow, at a pressure of 30 mbar, the image was taken 500 µs after the start of the expansion.

In Figure 36 (b), it is possible to see greater sharpness in the contours of the flow, indicating that part of the light has been absorbed by the iodine seeds, thus causing an increase in contrast in the image. Figures 37 and 38 show the average speed of the molecules as a function of distance. Measurements were taken at 30, 20 and 15 mbar of internal

chamber pressure.

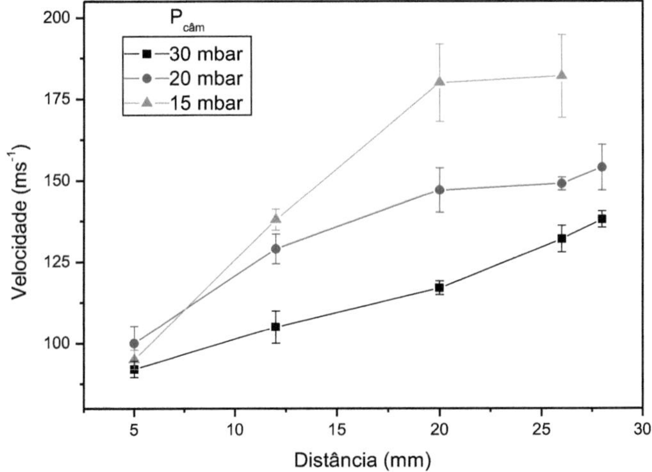

Figure 37 - Average flow velocity of pure SF₆ .

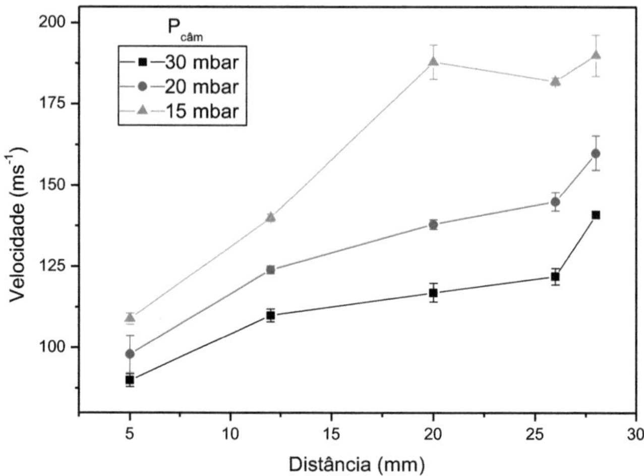

Figure 38 - Average flow velocity of SF₆ with iodine seeds.

According to the results in Figures 37 and 38, there is no appreciable difference in the distribution of velocities throughout the flow. This indicates that, in the velocity range investigated, the iodine seeds do not interfere with the flow velocity. The maximum average flow velocity is approximately 200 ms^{-1} , at a chamber pressure of 15 mbar.

In addition to the study at lower pressures, about an order of magnitude compared to the use of the LED, the laser made it possible to analyse the flow in regions up to 30 mm further from the valve, as opposed to 10 mm with the LED. In addition, it can be seen that the error in the measurements is smaller in the iodised flow (Table 8). The greater precision in the velocity measurements was due to better visualisation of the flow contours.

Table 8 shows the Mach number and temperature calculated for the flows in Figure 38, for pressures of 20 mbar. The method used for the calculation was the same as that used in the schlieren velocimetry and FIG. Figure 39 represents the data from Table 8 for the Mach number.

Table 8 - Mach number and temperature.

x (mm)	M	T(K)	M_{I2}	T_{I2} (K)
5	0,306±0,036	294	0,299±0,018	295
12	0,397±0,007	291	0,381±0,003	292
20	0,454±0,010	288	0,425±0,005	290
26	0,461±0,018	288	0,448±0,009	288
28	0,477±0,031	287	0,496±0,016	286

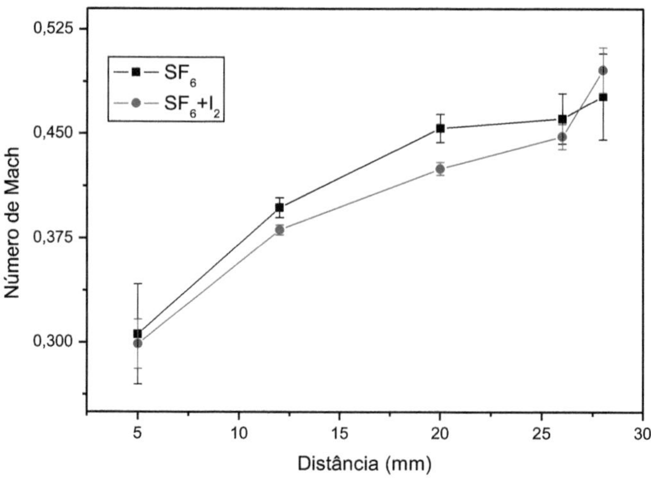

Figure 39 - Mach number of the pure flow with iodine seeds.

For pulses with a valve opening time of 20 µs, the maximum Mach number was approximately 0.5 and the minimum temperature was approximately 286 K at a distance of 28 mm. The study of situations closer to supersonic flow comes up against two problems: the intensity of iodine absorption, which decays with temperature, and the iodine seeding process in the gas.

In fact, the results showed that the seeding of the iodine in the SF_6 gas is not uniform. When it sublimates in the ampoule, the iodine is dragged along by the pressurised SF_6 gas. Some of the sublimated iodine returns to a solid state as it passes through the polyethylene tube, solidifying on the tube walls. One way to avoid resublimation was to reduce the length of the tube. However, it was still not possible to achieve a uniform seeding distribution.

The seeding process is essential for reproducibility in increasing contrast. The method used for seeding should be changed in future work. One proposal is to use nitrogen dioxide molecules, NO_2 , to seed atmospheric gas flows. In Laser Induced Fluorescence velocimetry, NO_2 has been used as a molecular marker [,7374], due to the presence of a strong absorption band in the visible region [,7576]. Consequently, it becomes an excellent option for seeding the flow, as it has an absorption band in the region of the lasers available in the laboratory and is in the gaseous state at room temperature.

5   Conclusion

Using the Rapid Ionisation Detector, it was possible to study flows produced at a pressure of 10^{-6} mbar and with velocities ranging from 21 to 726 ms^{-1}. Using the characteristics method to calculate the temperature and Mach number, it was possible to obtain flows with a minimum temperature of 29 K and a maximum Mach number of 12. The standard deviation in the velocity measurements was a maximum of 3%. This result showed that the pulses produced under the same conditions are very similar. The results of the flow study using the Fast Ionisation Detector showed that it was possible to obtain supersonic flows in continuous and transition regimes. However, it was not possible to achieve a free molecular regime.

The full-field Schlieren optical array equipped with a high-speed camera proved to be a suitable technique for determining the velocity of flows produced in an expansion chamber. It was possible to visualise flows at pressures of up to 250 mbar and with velocities ranging from 5.2 to 53.5 ms^{-1}. The flow temperature was *ca.* 300 K, with no significant variation, while the Mach number ranged from 0.001 to 0.144.

By modifying the schlieren method, it was possible to visualise flows at pressures of up to 15 mbar, an order of magnitude lower than the conventional schlieren method. With the SCAM method, flow velocities ranging from 92 to 190 ms were determined^{-1}. The temperature ranged from 295.3 to 280.1 K and the maximum Mach number was 0.6. The SCAM method proved promising for applications in the study of flows in the hypersonic shock tunnel.

With the velocity data obtained from the three velocimetry methods, it was possible to characterise the flows produced from the expansion chamber, creating a database that will be used as a reference for establishing the other non-intrusive diagnostic techniques currently under investigation in our research group.

6 References

[1] ESTRUCH, D.; LAWSON, N. J.; GARRY, K. P. Application of Optical Measurement Techniques to Supersonic and Hypersonic Aerospace Flows. **J. Aerosp. Engineer,** v. 22, n. 4, p. 383-395, oct. 2009.

[2] MIELKE, A. F. et. al. Time-average measurement of velocity, density, temperature, and turbulence velocity fluctuations using Rayleigh and Mie scattering. **Experim. in Fluids,** v. 39, n. 2, p. 441-454, aug. 2005.

[3] OBERKAMPFA, W. L.; TRUCANO, T. G. Verification and validation in computational fluid dynamics. **Progr. in Aerosp. Scienc.,** v. 38, n. 3, p. 209-272, apr. 2002.

[4] KEGELMAN, J. T.; DANEHY, P. M.; SCHWART, R. J. **Advanced Capabilities for Wind Tunnel Testing in the 21st Century.** Orlando, FL. 2000, p. 1 (NASA TM- 3044).

[5] OLIVEIRA, A. C. **Experimental investigation of laser energy addition in low density hypersonic flow.** 2008. 202 f. Thesis (Doctorate in Space Engineering and Technology) - National Institute for Space Research, São José dos Campos.

[6] SANTOS, A. M. A pesquisa e desenvolvimento em hipersônica no IEAv. **Revista Brasileira de Aplicações de Vácuo,** v. 27, n. 1, p. 5-10, 2008.

[7] DO, H.; MUNGAL, M. G.; CAPPELLI, M. A. Jet Flame Ignition in a Supersonic Crossflow Using a Pulsed Nonequilibrium Plasma Discharge. **IEEE Transactions on Plasma Science,** v. 36, n. 6, p. 2918-2923, dec. 2008.

[8] ALFEROV, V. I.; BUSHMIN, A. S.; SHINELEV, A. A. Possible Uses of Optical Methods for Determining the Structure of Nonequilibrium Hypersonic Flows. **High Temperature,** v. 45, n.3, p. 377-383, jun. 2007.

[9] MIELKE, A. F.; ELAM, K. A. Dynamic measurement of temperature, velocity, and density in hot jets using Rayleigh scattering. **Experiments in Fluids,** v. 47, p. 673-688, oct. 2009.

[10] GUSTAVSSON, J. P. R.; SEGAL, C. Filtered Rayleigh scattering velocimetry accuracy investigation in a M=2.2 axisymmetric jet. **Experiments in Fluids**, v. 38, n. 1, p 11-20, jan. 2005.

[11] CRAFTON, J.; et. al. Filtered Rayleigh Scattering Velocimetry for Wind Tunnel Applications. Dayton, OH: Innovative Scientific Solutions Inc. Apr 2004. 188 p, (ADA426476).

[12] MOHAMED, A. K. et. al. Electron Beam Fluorescence in Hypersonic Facilities, **The ONERA Journal Aerospace Lab**, n. 1, p. 1-9, dec. 2009.

[13] Yang, P.; Seitzman J. M., Particle vaporisation velocimetry for soot-containing flows. In: AIAA AEROSPACE SCIENCES MEETING AND EXHIBIT, 38, 2000, Reno. **Proceedings AIAA 38th Aerospace Sciences Meeting and Exhibit: Reston,** AIAA 2000. p. 1-10.
[14]BARKER, P. et. al. Velocity measurements by flow tagging employing laser enhanced ionisation and laser induced fluorescence. **Spectrochimica Acta Part B,** v. 50, n. 11, p. 1301-1310, sep.1995.

[15] IFFA, E. D. A.; AZIZ, R. A.; MALIK, A. S. Velocity field measurement of around jet using quantitative schlieren. **Applied Optics,** v. 50, n. 5, p. 618-625, feb. 2011.

[16] SRIVASTAVA, A. Development and application of colour schlieren technique for investigation of three-dimensional concentration field. **Journal of Crystal Growth.** v. 383, n. 15, p. 131-139, nov. 2013.

[17] KASHITANI, M.; YAMAGUCHI, Y. Flow Visualisation around a Double Wedge Air foil Model with Focusing Schlieren System. **Journal of Thermal Science,** v.15, n.1, p. 31-36, mar. 2006.

[18] MEIER, A. H.; ROESGEN, T. Improved background oriented schlieren imaging using laser speckle illumination. **Exp. Fluids,** v. 54, p. 1-6, jun. 2013.

[19] SETTLES, G.S. **Schlieren and shadowgraph techniques:** visualising phenomena in transparent media. Berlin: Springer-Verlag. 2001, v. 1, 376 p.

[20] MERZKIRCH, W. F. Sensitivity of flow visualisation methods at low-density flow conditions. **AIAA Journal,** v. 3, n. 4, pp. 794-795, apr. 1965.

[21] RIENITZ, J. Schlieren experiment 300 years ago. **Nature,** v. 254, p. 293-295, mar. 1975.

[22] MAZUMDAR, A. **Principles and techniques of schlieren imaging systems.** New York, NY: Columbia University, 2013. 16 p. (CUCS-016-13).

[23] OLIVEIRA, A. C. et. al. Bow Shock Wave Mitigation by Laser-Plasma Energy Addition in Hypersonic Flow. **Journal of Spacecraft and Rockets,** v. 45, p. 921-927, sep-oct. 2008.

[24] ROLIM, Tiago Cavalcanti et. al. Experimental results of Mach 10 conical-flow derived waverider to 14-X hypersonic aerospace vehicle. **Journal of Aerospace Technology and Management**, v. 3, p. 127-136, May-Aug. 2011.

[25] TORO, P. G. P. et. al. Experimental Hypersonic Investigation over the Micro-Satellite SARA. In: National Meeting of Thermal Sciences, 2004, Rio de Janeiro. **Proceedings of the 10th Brazilian Congress of Thermal Sciences and Engineering - ENCIT,** 2004.

[26] SALVADOR, Israel Irone et. al. Hypersonic Experimental Analysis of Impulse Generation in Airbreathing Laser Thermal Propulsion. **Journal of Propulsion and Power (Print),** v. 1, p. 1-14, may-jun. 2013.

[27] GONZÁLEZ, R. S., at. al. Repetitively Pulsed Hypersonic Flow Apparatus for Diagnostic Development. **AIAA journal,** v. 50, n. 3, p. 691-697, mar. 2012.

[28] ANDREA G. H. et. al. Two-component molecular tagging velocimetry utilising NO fluorescence life time and NO2 photodissociation techniques in an under expanded jet flow field. **Applied optics,** v. 48, n. 22, p. 4414-4423, aug. 2009.

[29] DELCHAR, T. A. **Vacuum Physics and Techniques.** London: Springer, 1993. v. 6, 253 p.

[30] BASU R; NAGUIB, A. M.; KOOCHESFAHANI, M. M. Feasibility study of whole-field pressure measurements in gas flows: molecular tagging manometry. **Exp. Fluids,** v. 49, n. 1, p. 67-75, jul. 2010.

[31] GONZÁLEZ, R. S.; BOWERSOX, R. D. W.; NORTH, S. W. Simultaneous velocity and temperature measurements in gaseous flow fields using the vibrationally excited nitric oxide monitoring technique: a

comprehensive study. **Applied optics,** v. 51, n. 9, p. 1216-1228, mar. 2012.

[32] CERCIGNANI, C. **Rarefied Gas Dynamics: From Basic Concepts to Actual Calculations**, New York: Cambridge University Press 2000, v.1, 327 p.

[33] KALEMPA, D. **Transport phenomena in the flow of rarefied gas mixtures**, 2005, 94 f. Dissertation (Master's in Physics) - Federal University of Paraná, Curitiba.

[34] QUINTELLA, C. M. Supersonic molecular beams in chemistry. **Química Nova,** v. 19, n. 6, p. 660-667, nov-dec.1996.

[35] KANTROWITZAND, A.; GREY J. A High Intensity Source for the Molecular Beam. Part I. **Review of Scientific Instruments**, v. 22, n. 5, p. 328-332, may. 1951.

[36] KISTIAKOWSKYAND, G. B.; WILLIAM, P. A High Intensity Source for the Molecular Beam. Part II. Experimental. **Review of Scientific Instruments**, v. 22, n. 5, p. 333-337. May. 1951.

[37] MORSE M. D. Supersonic Beam Sources, **Experimental Methods in the Physical Sciences, Atomic, Molecular, and Optical Physics**, v. 29, Part B, p. 21-47, 1996.

[38] ANDERSON J. B.; FENN, J. B. Velocity Distributions in Molecular Beams from Nozzle Sources, **Physics of Fluids**, v.8, n. 780, p. 1958-1988, may. 1965.

[39] RAMSEY, N. **Molecular Beams**, London: Oxford University Press, 1956, v. 1, 466 p.
[40] NUSSENZVEIG, H. M. **Curso de Física Básica 2:** Fluidos, Oscilações e Ondas Calor. São Paulo: Edgard Blucher, 1996, v. 2, p. 122-143.

[41] HOLLAS, J. M.; PHILLIPS, D. **Jet Spectroscopy and Molecular Dynamics**. Glasgow: Springer, 1995. v. 1, 446 p.

[42] SALEH, H. J.; MCCAFFERY, A. J. Alignment of Diatomic Molecules in a Free-jet Expansion. **J. chem. Doc. Faraday Trans.**, v. 89, n. 17, p. 3217-3221, 1993.

[43] YANG, S.; DAINEKA, D. V.; CHÂTELET, M. Experimental investigations of size distribution through large van der Waals cluster beam cross-section. **Chemical Physics Letters**, v. 377, p. 595-600, aug. 2003.

[44] GENTRYAND, W. R.; GIESE, C. F. Ten-microsecond pulsed molecular beam source and a fast ionisation detector. **Review of Scientific Instruments**, v. 49, n. 1, p. 595-600, may. 1978.

[45]CROSS, J. B.; VALENTINI, J. J. High repetition rate pulsed nozzle beam source, **Review of Scientific Instruments**, v. 53, n. 1, p. 38-42, sep. 1982.

[46] YAN, B. et. al. A new high intensity and short-pulse molecular beam valve. **Review of Scientific Instruments**, v. 84, p. 0231021-0231028, feb. 2013.

[47] ZMIJANOVIC, V. et. al. Experimental and Numerical Study of Thrust-Vectoring Effects by Transverse Gas Injection into a Propulsive Axisymmetric C-D Nozzle. In. 48th AIAA/ASME/SAE/ASEE Joint Propulsion Conference & Exhibit, 48, 2012, Atlanta. **Proceedings AIAA**, Reston, p. 1-16.

[48] TRAVA-AIROLDI, V. J. **Interaction of Intense Electromagnetic Fields with SF6 Molecules and their Agglomerates in Supersonic Expansion**. 1986. Thesis (Doctorate in Physics) - Technological Institute of Aeronautics, São José dos Campos.

[49] WILLEMS, P.; HULSMAN, H.; AERTS, F. On the use of laser-induced fluorescence to study free-jet expansions, **Chemical Physics**, v. 71, n. 1, p. 27-39 sep. 1982.

[50] SRETENOVIĆ, G. B. et al, Spatio-temporally resolved electric field measurements in helium plasma jet. **Journal of Physics D: Applied Physics**, v. 47, n. 10, p. 1-7, feb. 2014.

[51] OWEN, P. L.; THORNHILL, C. K. **The Flow in an Axially-Symmetric Supersonic Jet from a Nearly-Sonic Orifice into a Vacuum**. London: Aeronautic. Res. Council, 1948, 10 p. 10. (RM 1626).

[52] ASHKENAS, H.; SHERMAN, F.S. In 4th International Symposium on Rarefield Gas Dynamics, Toronto, 1964. **Proceedings of the Fourth International Symposium on Rarefied Gas Dynamics,** New York, 1965.

[53] CASSANOVA, R. A.; STEPHENSON, W. B. Expansion of a jet into a vacuum. **Symposium International on Combustion**, v. 11, n. 1, p. 577-587, 1967.

[54] TOLEDO, A. O. **Caracterização de Jatos Moleculares Supersônicos por Espectroscopia de Massa e Espectroscopia de Alta Resolução**. 1996, 89 p. Thesis (Doctorate in Physics) - Instituto Tecnológico de Aeronáutica, São José dos Campos.

[55] ANDERSON, J. D. **Modern Compressible Flow: With Historical Perspective**. Singapore: Mcgraw-hill College, 1990, v.1, 650 p.

[56] LASERTECHNICS, INC. *Model 203B pulsed valve manual*, Albuquerque, NE. 1984.

[57] BEAM DYNAMICS. *Rapid Ionisation Detector Manual, model FIG 1*. Minneapolis, MN, 17 p.
[58] TOLEDO, A. O. ; ANTUNES, L. M. D. ; SBAMPATO, M. E. ; SANTOS, A. M. . SF_6 molecule temperature in free jets obtained by diode laser spectroscopy and gasdynamics methods. in: Simposium on lasers and their applications - sla97, 1997, Campinas - SP. **Proceedings of the Simposium on lasers and their applications**, 1997. v. único, p. 380-383.

[59] TRAVA-AIROLDI, V. J.; SANTOS, R. Velocity measurements of a gas in supersonic flow using a laser beam. **Revista Brasileira de Aplicações de Vácuo**, v. 2, n.2, p. 209-215, 1982.

[60] TRAVA-AIROLDI, V. J. et. al. Characterisation of free jet expansion of SF_6 molecules. **Journal of Applied Physics**, v. 61, n.7, p. 2674-2676, 1987.

[61] ALTUCCI, C, at. al. Characterisation of pulsed gas sources for intense laser field-atom interaction experiments. **J. Phys. D: Appl. Phys.** v. 29, p. 68-75, jun 1996.

[62] P. ANDRESEN et. al. Characteristics of a piezoelectric pulsed nozzle beam. **Review of Scientific Instruments,** v. 56, p. 2038-2042, apr.1985.

[63] IZAWA1, M., KITA1, S.; INOUYE1, H. Some characteristics of pulsed nozzle beams. **J. Appl. Phys.** v. 53, n. 7, p. 4688-4694, jul. 1982.

[64] VRAKER, W.; MOSSMAN, A. L. **Gas Data Book.** New Jersey: Matheson, 1980, v.1, p.

[65] SEASHOLTZ, R. G.; BUGGELE, A E.; REEDER, M. F. Flow measurements based on Rayleigh scattering and Fabry-Perot interferometer. **Optics and Lasers in Engineering,** v. 27, n. 6, p. 543-570, Aug. 1997.

[66] SEASHOLTZ, R. G.;BUGGELE, A. E. **Improvement in Suppression of Pulsed Nd:YAG Laser Light With Iodine Absorption Cells for Filtered Rayleigh Scattering Measurements.** Cleveland, OH: NASA, Lewis Research Centre, 1997. NASA TM-113177.

[67] OSWALDO, S. I_2 - A didactic molecule. **Chem. Nova,** v. 31, n. 4, p. 914-920, mar. 2008.

[68] SALAMI, H.; ROSS, A. J. A molecular iodine atlas in ascii format. **Journal of Molecular Spectroscopy,** v. 233, n. 1, p. 157-159, sep. 2005.

[69] ELORANTA, E. W.; RAZENKOV, I. A. Frequency locking to the centre of a 532 nm iodine absorption line by using stimulated Brillouin scattering from a single-mode fibre. **Optics Letters,** v. 31, n. 5, p. 598-600, mar. 2006.

[70] KNOCKEL, H; TIEMANN, E. **Iodine Spectrum Calculating Software: IodineSpec,** for Windows: version 5.0:[S1], Hannover, 2011, Toptica Photonics, 1 CD-ROM.

[71] GERSTENKORN, S.; LUC, P.; PERRIN, A. Rotational analysis of the 5350 Å band of iodine by means of Fourier transform spectroscopy. **Journal of Molecular Spectroscopy,** v. 64, n.1, p. 56-69, jan. 1977.
[72] LUC, P. Molecular Constants and Dun ham Expansion Parameters Describing the B-X System of the Iodine Molecule. **Journal of Molecular Spectroscopy.** v. 80, n. 1, p. 41-55, mar. 1980.

[73] ELBAZ, A. M.; PITZ, R.W. N_2 O molecular tagging velocimetry. **Appl. Phys. B,** v. 106, p. 961-969, feb. 2012.

[74] JIANG, N.; NISHIHARA, M.; LEMPERT, W. R. Quantitative NO_2 molecular tagging velocimetry at 500 kHz frame rate. **Appl. Phys. Lett.** v. 97, n. 22, p. 2211031-2211033, nov. 2010.

[75] A.C. VANDAELE, Absorption cross-sections of NO_2 : simulation of temperature and pressure effects. **Journal of Quantitative Spectroscopy & Radiative Transfer**, v. 76, n. 3, p. 373-391, feb. 2003.

[76] COQUART, B.; JENOUVRIER, A.; MERIENNE, M. F. The NO_2 absorption spectrum. II: absorption cross-sections at low temperatures in the 400-500 nm region. **Journal of Atmospheric Chemistry**, v.21, n. 3, p. 251-261, jul. 1995.

Appendix A: Photos of the equipment used in the experiments

Vacuum chamber

Pulsed valve

Valve system

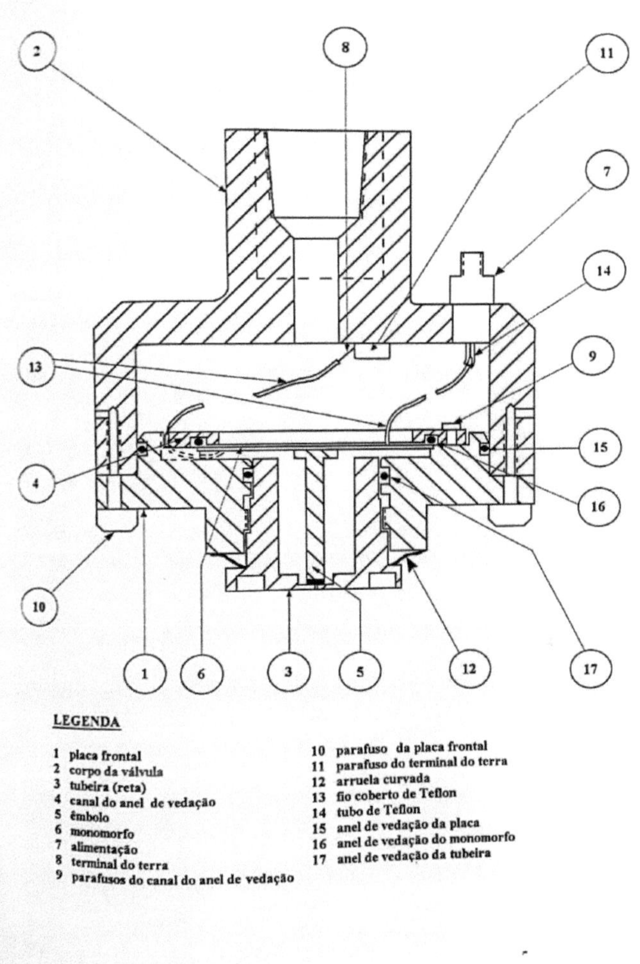

Pulsed valve driver

Wave function generator

Rapid ionisation detector

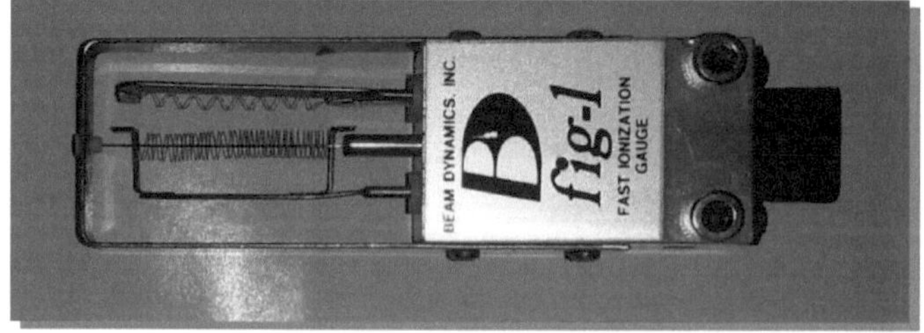

Fast ionisation detector driver

Oscilloscope

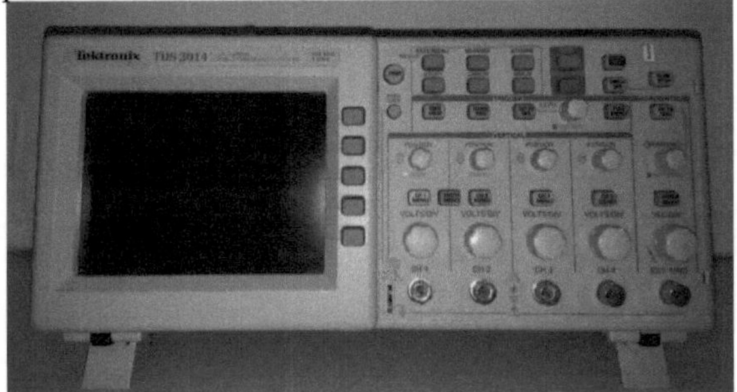

Schlieren optical arrangement

Schlieren optical arrangement

Schlieren Optical Arrangement Combined with Molecular Absorption

Schlieren Optical Arrangement Combined with Molecular Absorption

Schlieren Optical Arrangement Combined with Molecular Absorption

Annex B: Speed tables for the Rapid Ionisation Detector

Valve opening time in 100 µs.

x (mm)	u [2 bar] (ms-1)	T [2bar] (K)	M [2 bar]	u [4 bar] (ms-1)	T [4 bar] (K)	M [4 bar]
6				63	298	0,19
16	153	287	0,47	150	288	0,46
26	201	278	0,63	225	272	0,72
36	240	268	0,77			
46				369	225	1,29
56				350	233	1,20
67	475	176	1,87	476	175	1,88
77				531	145	2,31
87						
97	675	49	5,06	640	75	3,88
117	629	82	3,63	612	94	3,31
127	657	62	4,37	618	90	3,41
137	681	45	5,32			
147				700	30	6,65
157						
167	723	13	10,70	703	28	6,99

Valve opening time in 100 μs.

x (mm)	u [6 bar] (ms-1)	T [6 bar] (K)	M [6 bar]	u [8 bar] (ms-1)	T [8 bar] (K)	M [8 bar]
6	72	297	0,22			
16	158	286	0,49	155	287	0,48
26	235	270	0,75	234	270	0,75
36				326	242	1,10
46	390	216	1,39	365	227	1,27
56	344	235	1,17	347	234	1,19
67	460	184	1,78	460	184	1,78
77	529	146	2,29	555	131	2,54
87	572	120	2,73	587	110	2,93
97	615	92	3,37	617	91	3,40
117	613	93	3,33			
127	643	73	3,95	623	87	3,50
137				667	55	4,70
147	722	13	10,38	684	42	5,51
157	695	35	6,20	699	31	6,56
167	726	10	12,02	715	19	8,62

Valve opening time in 150 μs.

x (mm)	u [2 bar] (ms-1)	T [2bar] (K)	M [2 bar]	u [4 bar] (ms-1)	T [4 bar] (K)	M [4 bar]
6	58	298	0,18	57	298	0,17
16	121	292	0,37	115	293	0,35
26	157	286	0,49	179	282	0,56
36	243	268	0,78			

x (mm)	u (ms⁻¹)	T (K)	M	u (ms⁻¹)	T (K)	M
46	287	255	0,94	300	250	0,99
56	380	220	1,34	412	207	1,50
67	400	212	1,44	406	210	1,47
77	481	173	1,92			
87	584	112	2,89	522	150	2,23
97	626	85	3,57	563	125	2,64
117	617	91	3,40	568	123	2,69
127	634	79	3,74	546	136	2,46
137	670	53	4,83	617	91	3,40
147	589	109	2,96	701	30	6,71
157				700	31	6,61
167	720	15	9,74	687	41	5,65

Valve opening time in 150 µs.

x (mm)	u [6 bar] (ms-1)	T [6 bar] (K)	M [6 bar]	u [8 bar] (ms-1)	T [8 bar] (K)	M [8 bar]
6	64	298	0,20	62	298	0,19
16				127	291	0,39
26	211	276	0,67			
36	255	264	0,82	265	261	0,86
46	292	253	0,96	299	251	0,99
56				309	247	1,03
67	401	211	1,45	396	214	1,42
77	459	184	1,78	465	181	1,81
87	530	145	2,31	481	173	1,92
97	551	133	2,51	537	141	2,37
117	567	123	2,68	559	128	2,59
127	599	103	3,10	567	123	2,68
137	587	110	2,93	572	120	2,74
147	668	54	4,75	633	80	3,71
157	672	52	4,91	668	55	4,72
167	632	81	3,69	690	38	5,86

Valve opening time in 200 µs.

x (mm)	u [2 bar] (ms-1)	T [2bar] (K)	M [2 bar]	u [4 bar] (ms-1)	T [4 bar] (K)	M [4 bar]
6	46	299	0,14	69	297	0,21
16	199	278	0,62			
26	170	284	0,53	182	282	0,57
36	237	269	0,76	237	269	0,76
46	289	254	0,95	300	250	0,99

x (mm)	u	T	M	u	T	M
56				357	230	1,23
67	406	209	1,47	394	215	1,41
77	457	185	1,76			
87	535	143	2,35	485	170	1,95
97				567	123	2,67
117						
127	648	69	4,09	553	132	2,53
137	625	85	3,55	617	91	3,40
147	683	43	5,44	659	61	4,41
157	679	46	5,23	687	40	5,69
167				666	56	4,66

Valve opening time in 200 μs.

x (mm)	u [6 bar] (ms-1)	T [6 bar] (K)	M [6 bar]	u [8 bar] (ms-1)	T [8 bar] (K)	M [8 bar]
6	68	297	0,21	68	297	0,21
16	118	292	0,36	119	292	0,36
26	200	278	0,63	199	278	0,63
36	243	268	0,78	244	267	0,78
46	272	259	0,89	289	254	0,95
56	344	235	1,17	309	247	1,03
67	381	220	1,34	378	221	1,33
77	448	190	1,70	427	200	1,59
87	493	166	2,01			
97	527	147	2,28	521	151	2,22
117	571	120	2,73	522	150	2,24
127				556	130	2,56
137	587	110	2,93	572	120	2,74
147	665	57	4,61	616	92	3,37
157	671	52	4,86			
167	659	61	4,42	642	73	3,94

Valve opening time in 250 μs.

x (mm)	u [2 bar] (ms-1)	T [2bar] (K)	M [2 bar]	u [4 bar] (ms-1)	T [4 bar] (K)	M [4 bar]
6				60	298	0,18
16				111	293	0,34
26	181	282	0,56	188	280	0,59
36	265	261	0,86	245	267	0,79
46	321	243	1,08			
56	356	230	1,23	297	251	0,98

67	408	209	1,48	395	214	1,42
77	444	192	1,68	454	187	1,74
87	537	141	2,37			
97	579	115	2,83			
117	591	108	2,98	566	124	2,66
127	631	81	3,68	552	132	2,52
137	584	112	2,89	554	131	2,54
147	676	49	5,06	654	64	4,28
157	672	52	4,88	644	72	3,98
167	655	64	4,30	659	61	4,40

Valve opening time in 250 μs.

x (mm)	u [6 bar] (ms-1)	T [6 bar] (K)	M [6 bar]	u [8 bar] (ms-1)	T [8 bar] (K)	M [8 bar]
6	68	297	0,21	62	298	0,19
16	126	291	0,39	123	292	0,38
26	199	278	0,62	191	280	0,60
36	266	261	0,86	256	264	0,83
46	287	255	0,94	273	259	0,89
56	337	238	1,14	315	245	1,05
67	376	222	1,32	371	224	1,30
77	453	187	1,74	442	193	1,67
87	493	167	2,00	480	173	1,91
97	532	144	2,32	523	150	2,24
117	552	132	2,51	523	149	2,24
127				560	128	2,60
137				545	137	2,44
147	659	61	4,41	616	92	3,37
157	648	69	4,08	649	69	4,11
167				619	90	3,43

Valve opening time in 300 μs.

x (mm)	u [2 bar] (ms-1)	T [2bar] (K)	M [2 bar]	u [4 bar] (ms-1)	T [4 bar] (K)	M [4 bar]
6	33	299	0,10	41	299	0,12
16	83	296	0,25	37	299	0,11
26	115	293	0,35	36	299	0,11
36	173	284	0,54			
46	221	273	0,70	219	274	0,69
56	262	262	0,85	248	266	0,80
67	294	253	0,97			

77	332	239	1,13	345	234	1,18
87	394	215	1,41	385	218	1,37
97				408	209	1,48
117	476	176	1,88	441	193	1,66
127				443	192	1,67
137	475	176	1,88	434	197	1,62
147				517	153	2,19
157				515	154	2,18
167	539	140	2,38	532	144	2,32

Valve opening time in 300 µs.

x (mm)	u [6 bar] (ms-1)	T [6 bar] (K)	M [6 bar]	u [8 bar] (ms-1)	T [8 bar] (K)	M [8 bar]
6	40	299	0,12	38	299	0,11
16	21	300	0,06	19	300	0,06
26	141	289	0,43	158	286	0,49
36	179	282	0,56	186	281	0,58
46	224	272	0,71			
56	289	254	0,95	251	265	0,81
67	288	255	0,95	284	256	0,93
77	336	238	1,14	330	240	1,12
87	382	220	1,35	360	229	1,25
97	420	203	1,55	393	215	1,40
117	425	201	1,57	417	204	1,53
127	457	185	1,76	442	192	1,67
137	483	171	1,94			
147	519	152	2,21	493	166	2,01
157	527	147	2,28	522	150	2,23
167	546	136	2,45			

Appendix C: Schlieren speed tables

Stagnation pressure of 1.5 bar and chamber pressure of 1 bar, valve opening time of 2.0 and 1.5 µs.

x (mm)	v [2 µs] (ms)-1	T [2 µs] K	M [2 µs]	v [1.5 µs] (ms)-1	T [1.5 µs] K	M [1.5 µs]

2	8,5	300,0	0,026	9,8	299,9	0,030
5	5,7	300,0	0,017	7,7	300,0	0,023
7	6,2	300,0	0,019	6,4	300,0	0,019
9	6,0	300,0	0,018	5,2	300,0	0,016

Stagnation pressure of 2 bar and chamber pressure of 1 bar, valve opening time of 2.0 and 1.5 μs.

x (mm)	v [2 μs] (ms)$^{-1}$	T [2 μs] K	M [2 μs]	v [1.5 μs] (ms)$^{-1}$	T [1.5 μs] K	M [1.5 μs]
2	15,1	299,9	0,046	17,3	299,8	0,052
5	10,5	299,9	0,032	11,2	299,9	0,034
7	8,6	300,0	0,026	9,1	300,0	0,028
9	7,7	300,0	0,023	8,8	300,0	0,027

[a] Stagnation pressure of 2 bar and chamber pressure of 1 bar, valve opening time of 1.0 μs. [b] Stagnation pressure of 3 bar and chamber pressure of 1 bar, valve opening time of 2.0 μs.

x (mm)	v [a] (ms)$^{-1}$	T [a] K	M [a]	v [b] (ms)$^{-1}$	T [b] K	M [b]
2	20,1	299,8	0,061	20,1	299,8	0,061
5	15,0	299,9	0,046	17,1	299,8	0,052
7	11,3	299,9	0,034	12,7	299,9	0,039
9	8,9	300,0	0,027	12,4	299,9	0,037

[a] Stagnation pressure of 3 bar and chamber pressure of 1 bar, valve opening time of 1.5 μs. [b] Stagnation pressure of 3 bar and chamber pressure of 1 bar, valve opening time of 1.0 μs.

x (mm)	v [a] (ms)$^{-1}$	T [a] K	M [a]	v [b] (ms)$^{-1}$	T [b] K	M [b]
2	15,1	299,9	0,046	17,3	299,8	0,052

5	12,5	299,9	0,038	12,7	299,9	0,039
7	10,8	299,9	0,033	9,8	299,9	0,030
9	8,9	300,0	0,027	8,7	300,0	0,026

Stagnation pressure of 3 bar and chamber pressure of 250 bar, valve opening time of 2.0 µs.

x (mm)	v (ms)$^{-1}$	T (K)	M
4	41,1	299,1	0,125
7	48,3	298,7	0,147
10	48,5	298,7	0,147
14	47,5	298,8	0,144

Stagnation pressure of 3 bar and chamber pressure of 250 bar, valve opening time of 2.0 µs.

x (mm)	v (ms)$^{-1}$	T (K)	M
5	33,9	299,4	0,103
8	41,9	299,0	0,127
11	45,6	298,9	0,138
14	46,8	298,8	0,142

Stagnation pressure of 3 bar and chamber pressure of 250 bar, valve opening time of 1.0 µs.

x (mm)	v (ms)$^{-1}$	T (K)	M
4	51,3	298,6	0,156
7	53,5	298,4	0,162
11	52,7	298,5	0,160
14	51,4	298,5	0,156

Stagnation pressure of 3 bar and chamber pressure of 300 bar, valve opening time of 2.0 µs.

x (mm)	v (ms)$^{-1}$	T (K)	M

4	48,3	298,7	0,147
7	45,3	298,9	0,137
9	45,0	298,9	0,137
13	42,8	299,0	0,130

Stagnation pressure of 3 bar and chamber pressure of 300 bar, valve opening time of 1.5 µs.

x (mm)	v (ms)$^{-1}$	T (K)	M
4	41,0	299,1	0,124
6	42,3	299,0	0,128
9	38,6	299,2	0,117
14	35,4	299,3	0,107

Stagnation pressure of 3 bar and chamber pressure of 300 bar, valve opening time of 1.0 µs.

x (mm)	v (ms)$^{-1}$	T (K)	M
3	48,3	298,7	0,147
6	43,1	299,0	0,131
8	44,3	298,9	0,134
12	43,4	299,0	0,132

Stagnation pressure of 3 bar and chamber pressure of 200 bar, valve opening time of 1.5 µs.

x (mm)	v (ms)$^{-1}$	T (K)	M
2	60,5	298,0	0,184
6	52,4	298,5	0,159
8	54,4	298,4	0,165
10	54,4	298,4	0,165

Stagnation pressure of 3 bar and chamber pressure of 200 bar, valve opening time of 1.0 µs.

x (mm)	v (ms)$^{-1}$	T (K)	M
3	66,4	297,6	0,202
5	66,5	297,6	0,202
8	62,1	297,9	0,189
11	53,8	298,4	0,163

Annex D: Sequence of images obtained using the SCAM technique

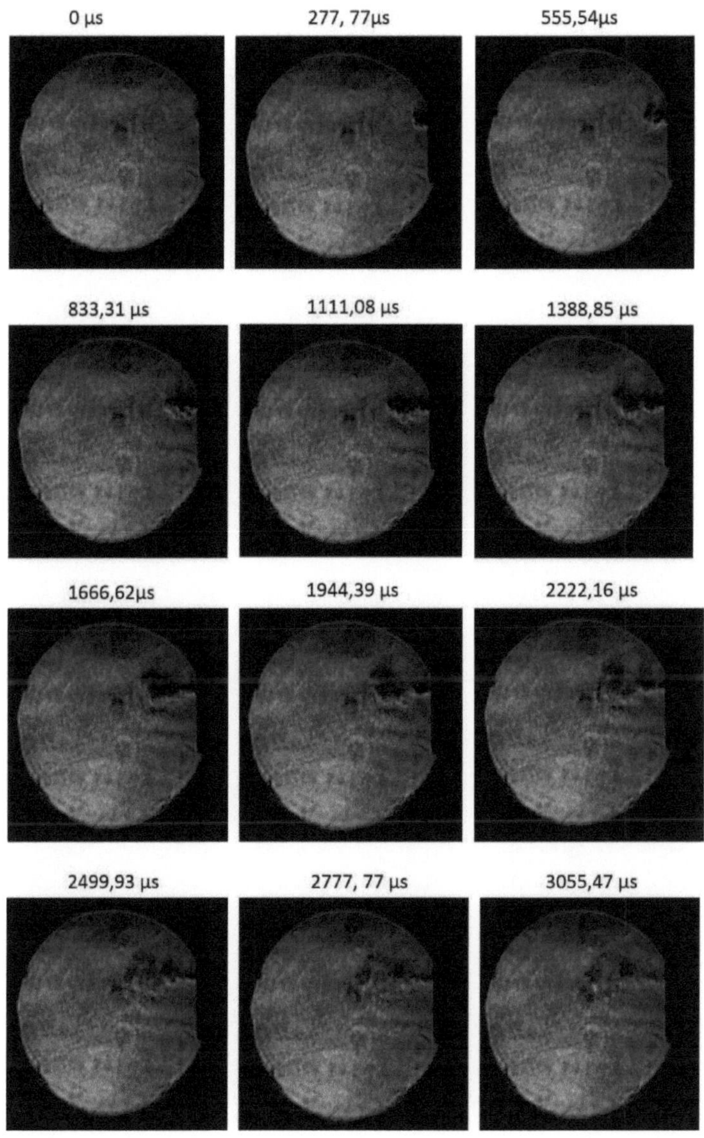

Buy your books fast and straightforward online - at one of world's fastest growing online book stores! Environmentally sound due to Print-on-Demand technologies.

Buy your books online at
www.morebooks.shop

Kaufen Sie Ihre Bücher schnell und unkompliziert online – auf einer der am schnellsten wachsenden Buchhandelsplattformen weltweit! Dank Print-On-Demand umwelt- und ressourcenschonend produzi ert.

Bücher schneller online kaufen
www.morebooks.shop

Printed by Books on Demand GmbH, Norderstedt / Germany